Primer for Data Analytics and Graduate Study in Statistics

Douglas Wolfe • Grant Schneider

Primer for Data Analytics and Graduate Study in Statistics

 Springer

Douglas Wolfe
Department of Statistics
Ohio State University
Columbus, OH, USA

Grant Schneider
Upstart Network, Inc.
Columbus, OH, USA

ISBN 978-3-030-47481-2   ISBN 978-3-030-47479-9   (eBook)
https://doi.org/10.1007/978-3-030-47479-9

This Springer imprint is published by the registered company Springer Nature Switzerland AG
The registered company address is: Gewerbestrasse 11, 6330 Cham, Switzerland

Douglas Wolfe: *To Robert V. Hogg and Allen T. Craig who fostered in me a passion for mathematical statistics during my graduate studies at the University of Iowa and to D. Ransom Whitney who nurtured my statistical career through my early years in academia at The Ohio State University.*

Grant Schneider: *To my parents, Bill and Judy Schneider, who nurtured my curiosity about the world by answering my countless questions as a kid, even when they didn't know the answers, and to Aunt Marce who gave the best, most clear-eyed advice a young man could receive, dispensed over warm Natty Lights.*

Preface

For many years, we taught a summer term course for the newly entering students in our graduate programs in Statistics at Ohio State University. It was designed to refresh and/or elevate the level of understanding for the basic background in probability and distributional theory required to be successful in our Master of Applied Statistics, M.S. in Statistics, and Ph. D. programs in Statistics and Biostatistics. Over the years, this proved to be an effective way for undergraduate students from a variety of quantitative backgrounds (particularly domestic students from smaller liberal arts programs) to bridge the transition from general undergraduate studies in a mathematically oriented field to the more career oriented graduate studies in Statistics.

Another factor that makes this text extremely relevant today is the recent increased interest in the field of data analytics (to be read "statistics" with a small s) as an undergraduate major. We recently started such a program here at Ohio State and within a few years it has attracted more than 100 top students. The job market is extremely strong for graduates with a data analytics undergraduate degree and it also provides an excellent background for those students who wish to continue their studies in a graduate program in statistics or biostatistics (where the job market is also outstanding).

We believe that this text provides the necessary framework for an undergraduate course at a smaller liberal arts college for anyone who is interested in either exploring job opportunities in the data analytics field itself or in attending a graduate program in statistics or biostatistics. It could also, of course, be used for a bridge course similar to the one we taught in the summer term to our incoming graduate students—or even as a good refresher text for a student who wishes to refresh/better prepare themselves for graduate work in statistics or biostatistics.

Columbus, OH, USA

Columbus, OH, USA

Douglas Wolfe

Grant Schneider

Contents

Chapter 1
Introduction

Have you ever played the game "Twenty Questions"? Typically, the first questions are used to determine whether the item of interest is a person, place, or thing. Everything in our world (or cosmos, for that matter) is one of these. Once the category is established, the questioning usually proceeds along the line of ascertaining physical or personal properties of the item, such as how big it is, whether it is alive, or whether it is famous, etc. All of these follow-up questions are designed to help the player understand more about the item of interest and, eventually, help the player correctly identify it.

The simple game of "Twenty Questions" is an example of what we all face in our everyday lives. We are constantly trying to learn more about and understand "items" that we encounter in our lives. "How long was that home run?", "How tall are you or how much do you weigh at a fixed point in time?", "How many single records did Elvis Presley sell?", "What was the average daily price of the Apple common stock over the past twelve months?", "How much did the light from a distant star bend when it passed a large exoplanet on its way to earth?", "How many calories are in a chocolate milkshake?", "How intensive was that recent earthquake or volcanic eruption?", "How much is the carbon dioxide from burning fossil fuels affecting the warming of our atmosphere and our oceans?", etc. Fortunately, mathematics provides the mechanism for addressing every one of these questions, not only in terms of the actual physical measurements required but also in terms of understanding the scientific concepts and structure that are critical for interpreting these measurements. We use it all the time to routinely measure lengths, areas, speeds, and distances. However, it also plays a major role in more complicated settings, such as determining the necessary rocket speed to put a satellite into earth orbit or to send a probe on an interstellar mission; understanding the bond structure of complicated molecules to facilitate chemical processes for synthesizing new compounds/materials for tackling human diseases; the development of new materials for constructing heat-resistant shields to enable safe space travel; and helping to understand the nature of "dark matter" and "dark energy" and their role in an ever-expanding universe.

© Springer Nature Switzerland AG 2020
D. Wolfe, G. Schneider, *Primer for Data Analytics and Graduate Study in Statistics*,
https://doi.org/10.1007/978-3-030-47479-9_1

Everything discussed in the previous paragraph relates to our attempts to measure (or interpret such measurements) *deterministic physical events* at a fixed point in time—and there are mathematical approaches to addressing such phenomena. However, many interesting phenomena do not fit into such a deterministic framework. For example, will a coin flip result in heads or tails (or even leaning up against the wall!)? Will a given treatment for lung cancer be effective at treating *my* lung cancer? Will the price of Apple common stock rise in tomorrow's market trading? How many miles per gallon will I get with my new car? What effect will a change of diet have on my high blood pressure? How does the amount of sleep I get affect my physical well-being? These questions all relate to *random (non-deterministic) events* that require a different analytical approach. However, once again mathematics comes to our rescue, as it also provides the necessary structure to facilitate the understanding of such random (i.e., uncertain) events. That area of inquiry is governed by the rules of probability, and we can use these rules to provide the framework in which to study random events.

The first part of this text is devoted to the development of these basic probability rules and discussion of how they can be used to better understand random events, including what to expect from repetitions of a random event and how to interpret the observed outcome of such repetitions. The second portion of the text incorporates these basic probability properties for random events into a more formal expanded structure, called probability distribution theory, that can be used to provide models for analyzing the outcomes from repetitions of random events. These models play a key role in interpreting data obtained from experimental repetitions through the use of statistical inference techniques such as point estimation, confidence intervals, and hypothesis tests, the discussion of which is left as the next intriguing subject for your further exploration!

Chapter 2
Basic Probability

2.1 Random Events and Probability Set Functions

What is the chance that you would win the national lottery Powerball if you buy a single ticket? If you buy 100 tickets? What proportion of individuals with acute myeloid leukemia will respond positively to the FDA-approved drug *cytarabine*? What percentage of women will develop some form of Alzheimer's disease during their lifetime? What percentage of men? What is the likelihood that the son of two left-handed parents will also be left-handed? How many touchdowns will be scored in all of the 2020 football bowl games? If you wanted to collect one US penny issued from the Philadelphia mint for each of the years from 1960 through 2019, how many pennies should you expect to examine before achieving this goal? What percentage of people believe that God plays a role in determining which team wins a sporting event?

All of these questions deal with issues of uncertainty, and the basic properties of probability and associated statistical methodology play an important role in helping us understand appropriate ways to address these questions. In this chapter, we introduce the concept of a random event and describe the basic probability rules associated with random events. These rules provide the necessary structure that underlies the more general theory of probability distributions and, eventually, statistical sampling and all of the methodology for statistical inference based on such samples.

> **Definition 2.1** A **random or chance experiment** is an experiment that can be repeated under "identical" conditions such that the outcome of any specific trial is not predetermined.

We are all familiar with random experiments: flip a coin; roll a six-sided die; deal a 13-card hand from an ordinary deck of 52 cards (without the jokers); crop yield in a

© Springer Nature Switzerland AG 2020
D. Wolfe, G. Schneider, *Primer for Data Analytics and Graduate Study in Statistics*,
https://doi.org/10.1007/978-3-030-47479-9_2

given year from planting a specific hybrid of corn; gas mileage from a specific model automobile; and survival time from a treatment for lung cancer.

> **Definition 2.2** The list of all possible outcomes for a random experiment is called the **sample space**—denoted by S.

Example 2.1 Flip a Coin
Here there are only two possible outcomes (if we rule out the possibility that the coin rolls and leans up against a wall!), heads or tails. Thus, we can denote the corresponding sample space by $S = \{$Head (H), Tail (T)$\}$.

Example 2.2 Roll a Six-Sided Die (with Usual Labels)
In this setting, there are six possible outcomes and the associated sample space is $S = \{1, 2, 3, 4, 5, 6\}$.

Example 2.3 Roll a Pair of Six-Sided Dice (with Usual Labels)
For this random experiment, there are several possible sample spaces that could be considered, depending on what is of interest for the outcome. Examples include:

(a) $S = \{(1, 1), (1, 2), \ldots, (6, 5), (6, 6)\}$, where the first integer represents the outcome of the first die and the second integer represents the outcome of the second die
(b) $S = \{2, 3, \ldots, 11, 12\}$, corresponding to the sum of the numbers on the two dice
(c) $S = \{$even, odd$\}$, if we are only interested in whether the sum of the numbers on the two dice is an even or odd number

Example 2.4 Toss a Coin Until the First Head Is Observed
The outcome of interest here is how many tosses are required to obtain this first head. If we include the toss on which the first head is observed, the associated sample space is given by

$$S = \{1, 2, 3, \ldots\ldots\}.$$

If we are only interested in how many tails occur **before** the first head, then the appropriate sample space is

$$S = \{0, 1, 2, \ldots\ldots\}.$$

In either case, we have clearly encountered our first infinite (countably so) sample space!

Example 2.5 Crop Yield in Bushels of Corn per Acre
Here we encounter a bit of a problem in specifying the appropriate sample space. Crop yield is clearly a continuous quantity (unlike the first four examples where the outcomes of interest were discrete in nature). Here our sample space will have to be

an interval on the real line, for which the lower endpoint must be 0. However, there is no natural upper endpoint for the appropriate interval, especially in view of the continued modest overall increase in crop yield from improved farming techniques. What is usually done in such situations is to designate the sample space to be the nonnegative real line, so that $S = [0, \infty)$. Clearly, very large yields will not be attainable, but this can be addressed when we eventually assign probabilities to this set of outcomes.

Example 2.6 Survival Time (in Years) Following Medical Treatment for Lung Cancer

Once again, we are dealing with a continuous quantity, for which an interval sample space is required. The lower endpoint is also clearly 0 (although hopefully not actually included in the sample space). But what should be the upper endpoint for the sample space? Here, based on practical experience, it would be natural to consider the sample space to be something like $S = (0, 100]$. However, as we shall see later, for statistical analysis purposes, it is often taken to be $S = (0, \infty)$.

> **Definition 2.3** A **set function** is a function that assigns real numbers to every set in a collection of sets.

Example 2.7 Non-face Cards in a 52-Card Deck of Cards

Let $\Omega = \{$collection of 40 non-face cards in an ordinary deck of 52 cards$\}$. For every subset $A \subset \Omega$, define the set function:

$$Q(A) = [\text{total number of pips (spots) on the cards in } A].$$

Thus, for example, if $A_1 = \{$all diamonds in $\Omega\}$, then $Q(A_1) = 10 + 9 + \ldots + 2 + 1 = (10)(11)/2 = 55$. Similarly, for $A_2 = \{$all the eight cards$\}$, we have $Q(A_2) = 4(8) = 32$. What does this set function assign to the subset $A_3 = \{$odd numbered cards in $\Omega\}$?

Example 2.8 Set Function on the Interval [0, 100]

Let $\Omega = [0, 100]$ and, for every $A \subset \Omega$, define the set function $Q(\cdot)$ by $Q(A) = \int_A x^4 \, dx$.

Then, if $A = [0, 1]$, it follows that

$$Q(A) = Q([0, 1]) = \int_0^1 x^4 \, dx = \frac{x^5}{5} \Big|_{x=0}^{x=1} = \frac{1}{5}.$$

Similarly,

$$Q([0, 1] \cup \{15\} \cup [20, 30]) = \int_0^1 x^4 \, dx + \int_{15}^{15} x^4 \, dx + \int_{20}^{30} x^4 \, dx$$

$$= \frac{x^5}{5}\Big|_{x=0}^{x=1} + \frac{x^5}{5}\Big|_{x=15}^{x=15} + \frac{x^5}{5}\Big|_{x=20}^{x=30}$$

$$= 4{,}220{,}000.2$$

Definition 2.4 A set function $P(\cdot)$ defined on all the subsets (called **events**) of a sample space S is called a **probability set function** (or just **probability function**) if it satisfies:

(i) $P(C) \geq 0$ for all $C \subset S$
(ii) $P(S) = 1$
(iii) $P\left(\overset{\infty}{\underset{i=1}{\cup}} C_i\right) = \overset{\infty}{\underset{i=1}{\sum}} P(C_i)$ for all events $C_1, C_2, \ldots \subset S$ such that $C_i \cap C_j = \emptyset$ for all $i \neq j$, that is, mutually exclusive events $C_1, C_2, \ldots$

(**Note**: This also implies that $P\left(\overset{n}{\underset{i=1}{\cup}} C_i\right) = \overset{n}{\underset{i=1}{\sum}} P(C_i)$ for any finite collection of n events $C_1, C_2, \ldots, C_n \subset S$ such that $C_i \cap C_j = \emptyset$, $i \neq j = 1, \ldots, n$.)

Example 2.9 Drawing a Card at Random from a Deck of 52 Playing Cards
Draw one card at random from an ordinary deck of 52 playing cards and let $S = \{52$ individual cards$\}$. Define the probability set function $P(\cdot)$ by

$$P(\text{any of the individual 52 cards}) = 1/52.$$

Consider the three events: $C_1 = \{$red card$\}$, $C_2 = \{$jack or club$\}$, and $C_3 = \{$jack$\}$. Then we have $P(C_1) = \frac{1}{2}$, $P(C_2) = 16/52$, and $P(C_3) = 4/52$. Moreover,

$$P(C_1 \cap C_3) = P(\text{jack of diamonds or jack of hearts}) = 2/52.$$

Example 2.10 Roll a Pair of Six-Sided Fair Dice
Consider the sample space $S = \{(x, y): x, y = 1, \ldots, 6\}$ and the probability set function that assigns equal probability to each of the 36 possible outcomes; that is, $P(\text{any outcome in } S) = 1/36$. It follows that

$$P(\text{sum of the dice is seven}) = P((3, 4) \cup (4, 3) \cup (5, 2) \cup (2, 5) \cup (1, 6) \cup (6, 1))$$
$$= 6/36$$

and

$P(\text{sum of the dice is an even number}) = P(\text{sum is 2 or 4 or 6 or 8 or 10 or 12})$

$$= \sum_{x=2(2)12} P(\text{sum} = x)$$

$$= \frac{1}{36} + \frac{3}{36} + \frac{5}{36} + \frac{5}{36} + \frac{3}{36} + \frac{1}{36} = \frac{1}{2}.$$

2.2   Properties of Probability Functions

Let S be a sample space and let $P(\cdot)$ be a probability function defined on the subsets of S.

Lemma 2.1 Probability of Unions
For any subsets $A, B \subset S$, it follows that $P(A \cup B) = P(A) + P(B) - P(A \cap B)$.

Proof From property (iii) of Definition 2.4, we have

$$P(A \cup B) = P(A) + P(B \cap A^c) = P(A) + [P(B) - P(A \cap B)]$$
$$= P(A) + P(B) - P(A \cap B). \quad \blacksquare$$

This computation for the probability of unions is illustrated pictorially in the following Venn diagrams.

P(A∪B)   P(A)   P(B)   P(A∩B)

Lemma 2.2 Probability of Complements
Let $A \subset S$. Then $P(A^c) = 1 - P(A)$.

Proof Using properties (ii) and (iii) of Definition 2.4, we have
$$P(A \cup A^c) = P(S) = 1 = P(A) + P(A^c) \Rightarrow \text{result}. \quad \blacksquare$$

This computation for the probability of a complement is illustrated pictorially in the following Venn diagrams.

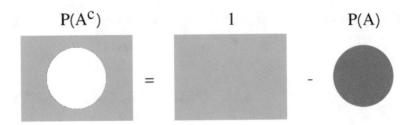

$$P(A^C) \qquad\qquad 1 \qquad\qquad P(A)$$

Example 2.11 Course Grades

Suppose that

$$P(A \text{ in Stat } 603) = .9, P(A \text{ in Stat } 602) = .8,$$

and $\qquad\qquad P(\text{at least one } A \text{ in Stat } 602 \text{ or } 603) = .95$

Then

$$P(A \text{ in both Stat } 602 \text{ and Stat } 603) = P(A \text{ in Stat } 602) + P(A \text{ in Stat } 603)$$
$$- P(A \text{ in at least one of them})$$
$$= .9 + .8 - .95 = .75;$$

$$P(A \text{ in neither course}) = 1 - P(A \text{ in at least one of them})$$
$$= 1 - .95 = .05;$$

$$P(A \text{ in Stat } 602 \text{ but not in Stat } 603) = P(A \text{ in Stat } 602) - P(A \text{ in both courses})$$
$$= .9 - .75 = .15;$$

and

$$P(\text{exactly one } A \text{ in the two courses}) = P(A \text{ in at least one of the courses})$$
$$- P(A \text{ in both courses})$$
$$= .95 - .75 = .20.$$

Keep in mind that there are often several ways to compute such probabilities. For example, we could alternatively have chosen to use the following approach:

$$P(\text{exactly one } A \text{ in the two courses})$$
$$= [P(A \text{ in Stat } 602) - P(A \text{ in both})] + [P(A \text{ in Stat } 603) - P(A \text{ in both})]$$
$$= (.9 - .75) + (.8 - .75) = .20.$$

> **Definition 2.5** Two events A and B are said to be **independent** if
> $$P(A \cap B) = P(A)\,P(B).$$

Note If A and B are independent events, then

$$P(A \cup B) = P(A) + P(B) - P(A)\,P(B).$$

Example 2.12 Which Player Should You Choose?

Consider playing the following game: A coin is flipped twice, with $p = P(\text{head})$. Player E has to give Player D \$5 if the two flips match, and Player D has to give Player E \$5 if one is a head and one is a tail. Which player do you want to be?

Let $\gamma = P(\text{both flips are the same})$. Then,

$$\begin{aligned}
\gamma &= P(\{H,H\} \text{ or } \{T,T\}) = P(\{H,H\}) + P(\{T,T\}) \quad \text{(disjoint events)} \\
&\underset{\text{ind}}{=} P(H \text{ on first})\,P(H \text{ on second}) + P(T \text{ on first})\,P(T \text{ on second}) \\
&= p^2 + (1-p)^2.
\end{aligned}$$

Using derivatives, we see that

$$\frac{d\gamma}{dp} = 2p - 2(1-p) = 4p - 2 = 0 \Rightarrow p = \frac{1}{2}.$$

Since $\left.\frac{d^2\gamma}{dp^2}\right|_{p=1/2} = 4 > 0$, we have that $\gamma = P(\text{both flips are the same})$ is minimized

when $p = \frac{1}{2}$, in which case, $\gamma = (.5)^2 + (1 - .5)^2 = .5$.

Thus, $P(\text{D wins}) = P(\text{E wins}) = .5$ when the coin is fair (i.e., $p = \frac{1}{2}$). However, $P(\text{D wins}) > P(\text{E wins})$ when the coin is not fair in any fashion. Has your opinion changed about which player you want to be?

Example 2.13 Drawing Marbles with Replacement Between Draws

Consider a box that contains 3 red marbles and 4 white marbles. We draw two balls at random from the box with replacement between the two draws. Then, with replacement between draws, the outcomes for the two draws are independent and we have

$$P(\text{both balls are red}) = \frac{3}{7}\left(\frac{3}{7}\right) = \frac{9}{49};$$

$$P(\text{both balls are white}) = \frac{4}{7}\left(\frac{4}{7}\right) = \frac{16}{49};$$

$$P(\text{both balls are the same color}) = \frac{9}{49} + \frac{16}{49}$$

$$= \frac{25}{49} > \frac{1}{2}. \quad (\text{Why can we just add them?})$$

Note This is the same as matching on two independent tosses of a coin with $p = 3/7$ or $4/7$.

Think About It How would $P(\text{both balls are the same color})$ change if we did not replace the ball between draws?

Example 2.14 Flip a Coin Until the First Head Occurs

Suppose we have a coin with $p = P(\text{head})$ and we flip it until the first head occurs. Let X denote the number of flips required. Then

(a) $P(X = 6) = P(\{T, T, T, T, T, H\})$

$$= P(\{T\})\, P(\{T\})\, P(\{T\})\, P(\{T\})\, P(\{T\})\, P(\{H\}) \quad (\text{Why?})$$

$$= (1 - p)^5 p;$$

(b) $P(\text{we need fewer than 4 flips}) = P(X < 4)$

$$= \sum_{x=1}^{3} P(X = x) = \sum_{x=1}^{3} p(1 - p)^{x-1}$$

$$= p + p(1 - p) + p(1 - p)^2$$

$$= p(3 - 3p + p^2);$$

(c) $P(\text{we need more than 2 flips}) = P(X > 2)$

$$= \sum_{x=3}^{\infty} P(X = x) = \sum_{x=3}^{\infty} p(1 - p)^{x-1}$$

$$= 1 - P(X \le 2) \quad (\text{Why?})$$

$$= 1 - \sum_{x=1}^{2} P(X = x)$$

$$= 1 - \sum_{x=1}^{2} p(1 - p)^{x-1}$$

$$= 1 - [p + p(1 - p)] = (1 - p)^2.$$

Think About It Does this result make intuitive sense?

2.3 Conditional Probability

Consider again the situation where we are drawing two marbles at random from a box that contains 3 red marbles and 4 white marbles. If we are interested only in the color of the second marble drawn, does it matter what color marble we obtained in the first draw? If the first marble is replaced before our second draw, then the color of the first marble drawn does not matter, and we know that the probability is 3/7 that the second drawn marble is red. What if we do not replace the first drawn marble before the second draw? Does the probability that the second drawn marble is red remain at 3/7 or does it increase or decrease depending on the color of the first drawn marble? Our intuition should suggest that without replacement between the draws, the probability for the second drawn marble being red is less than 3/7 or greater than 3/7 depending on whether the first drawn marble is red or white, respectively. Of course, we would want to be more specific about the exact nature of this change in probability, which leads us naturally to the concept of conditional probability.

Definition 2.6 Let A and B be any two events with $P(B) > 0$. Then the **conditional probability that A occurs given that B has occurred** is given by

$$P(A|B) = \frac{P(A \cap B)}{P(B)}.$$

Note: $P(A|B)$ is simply the proportion of $P(B)$ that also belongs in event A.

Example 2.15 Playing a Board Game
Consider a board game with a stack of three good cards and two penalty cards. Player A chooses a card first and then Player B chooses a card without A replacing her card. Which player has a greater chance of choosing a good card?

$P(\text{A chooses a good card}) = 3/5$

$P(\text{B chooses a good card}) = P(\text{A chooses a good card and B chooses a good card})$
$\qquad\qquad\qquad + P(\text{A chooses a penalty card and B chooses a good card})$
$\qquad\qquad = P(\text{A good}) \, P(\text{B good}|\text{A good})$
$\qquad\qquad\qquad + P(\text{A penalty}) \, P(\text{B good}|\text{A penalty})$
$\qquad\qquad = (3/5)(2/4) + (2/5)(3/4) = 3/5.$

Think About It Are these two probabilities affected if Player A returns her card to the deck and reshuffles before Player B draws a card?

Thus, **before the game begins**, the two players have an equal chance of choosing a good card on their first draws regardless of whether or not Player A returns her card to the deck and reshuffles the deck before Player B draws a card. Of course, if you obtain information from Player A's reaction to the card she drew, then the conditional probability that B draws a good card is either 2/4 or 3/4, depending on whether or not Player A was upset with the card she drew.

Example 2.16 Families with Two Children
Suppose a family has two children and we know that at least one of the children is a boy. Then

$$P(\text{first child is a boy}) = \frac{1}{2},$$

$$P(\text{second child is a boy}|\text{first child is a boy}) = \frac{1}{2},$$

$$P(\text{both children are boys}) = \left(\frac{1}{2}\right)\left(\frac{1}{2}\right) = \frac{1}{4}, \quad (\text{Why?})$$

and

$$P(\text{at least one child is a boy}) = 1 - P(\text{both children are girls}) = 1 - \frac{1}{4} = \frac{3}{4}.$$

Hence, it follows from our definition of conditional probability that

$$P(\text{both children are boys}|\text{at least one of the children is a boy})$$
$$= P(\text{both children are boys})/P(\text{at least one of the children is a boy})$$
$$= \frac{\frac{1}{4}}{\frac{3}{4}} = \frac{1}{3}.$$

Think About It Does your intuition agree with this result?

This raises the interesting question: does conditioning on an event always increase the probability of a second event, as it did in this example? Not necessarily, as the additional information from the conditioning on one event can either increase or decrease (or leave unchanged in the case of independent events) the probability of a second event.

Example 2.17 Rolling a Pair of Fair Dice
Roll a pair of fair dice and consider the following two events:

$$A = \{\text{sum is 6}\} \quad B = \{\text{sum is an even number}\}.$$

Then we have

$$P(A) = 5/36 \quad P(A^c) = P(\text{sum is not six}) = 31/36$$
$$P(B) = \tfrac{1}{2} \quad P(A \cap B) = P(A) = 5/36.$$

Now,

$$P(A|B) = \frac{P(A \cap B)}{P(B)} = \frac{5/36}{1/2} = \frac{5}{18} > P(A) = \frac{5}{36},$$

but

$$P(A^c|B) = \frac{P(A^c \cap B)}{P(B)} = \frac{P(2, 4, 8, 10, 12)}{1/2} = \frac{13/36}{1/2} = \frac{13}{18} < P(A^c) = \frac{31}{36}.$$

Comments

2.1 Conditional probabilities follow all the probability rules, just like unconditional probabilities. For example,

$$P(A^c|B) = 1 - P(A|B)$$

and

$$P(A \cup B|C) = P(A|C) + P(B|C) - P(A \cap B|C).$$

But be careful—don't get creative on the conditioning portion of a probability expression:

$$P(A|B) \neq 1 - P(A|B^c)$$

and

$$P(A|B \cup C) \neq P(A|B) + P(A|C), \quad \text{even if } B \cap C = \varnothing.$$

2.2 Disjoint events A and B *cannot* also be independent events unless at least one of them has zero probability.

Proof   Since A and B are independent events, then $P(A \cap B) = P(A)P(B)$. But A and B are also disjoint, so that $P(A \cap B) = P(\varnothing) = 0$, which implies that $P(A)P(B) = 0$, so that either $P(A) = 0$ or $P(B) = 0$ (or both, of course). ■

Lemma 2.3 Probability of Intersections: Multiplication Rule
Let A and B be events such that $P(A) > 0$ and $P(B) > 0$. Then

$$P(A \cap B) = P(A|B) \times P(B) = P(B|A) \times P(A).$$

Proof   This follows immediately from the definition of conditional probability. ■

Example 2.18
Consider again a box that contains 3 red marbles and 4 white marbles. In a previous example, we drew two marbles at random from the box with replacement between the two draws, and we found the following probabilities:

$P(\text{both marbles are red}) = (3/7)(3/7) = 9/49;$

$P(\text{both marbles are white}) = (4/7)(4/7) = 16/49;$

$P(\text{both marbles are the same color}) = (9/49) + (16/49) = (25/49) > 1/2.$

Note Remember that this is the same as matching on two tosses of a coin with $p = 3/7$ or $4/7$.

Now suppose we do *not* replace the first marble before we draw the second one. How will this affect these probabilities? Without replacement between draws, we see from the Multiplication Rule (Lemma 2.3) that

$P(\text{both marbles are red}) = P(\text{first is red})\, P(\text{second is red}|\text{first is red})$
$$= (3/7)(2/6) = 6/42;$$

$P(\text{both marbles are white}) = P(\text{first is white})\, P(\text{second is white}|\text{first is white})$
$$= (4/7)(3/6) = 12/42;$$

and

$P(\text{both marbles are the same color}) = P(\text{both white}) + P(\text{both red})$
$$= (6/42) + (12/42) = (18/42) < 1/2.$$

Think About It Does your intuition agree with these results? How does the probability that the two marbles are different colors compare under replacement between draws versus non-replacement between draws? Knowing when and how to condition matters!

Lemma 2.4 Extension of Multiplication Rule
Let A, B, and C be events. Then

$$P(A \cap B \cap C) = P(A|B \cap C)\, P(B|C)\, P(C).$$

This result extends further in a natural way to the intersection of an arbitrary number of events.

Example 2.19 Attending an Early History Lecture
We have an early history lecture at 8 am that we often consider excuses for not attending. We have a drawer that contains eight pairs of socks, each pair of a

different color. However, when we do our wash, we just throw all of the socks into the drawer (16 total) without pairing them by colors. At the suggestion of our statistics instructor (whose class is at 11 am), we have decided to use probability to determine if we should attend our 8 am history lecture tomorrow morning. When our alarm goes off, we will reach into the drawer *without looking* and draw a sock out at random—then we will draw a second sock out at random (without putting the first one back in!). If they match, we will go to our 8 am history lecture. If they do not match, we will draw a third sock (again without replacement between draws). If there is a match among these three socks, we will go to the history lecture. We continue this process up to five draws—if we do not have a match in these five draws, we will skip the history lecture and go back to bed. What is the probability that we will attend tomorrow's 8 am history lecture?

$$P(\text{attend the history lecture}) = P(\text{at least one color match in the first 5 draws})$$
$$= 1 - P(\text{no matches in the first 5 draws})$$
$$= 1 - \left(\frac{16}{16}\right)\left(\frac{14}{15}\right)\left(\frac{12}{14}\right)\left(\frac{10}{13}\right)\left(\frac{8}{12}\right)$$
$$= 1 - .4103 = .5897.$$

Note Our history instructor also liked the idea of the experiment, but she wanted us to continue to draw until the ninth sock had been drawn before we make a decision about going back to bed. . .

Example 2.20 Birthday Problem
Consider a class of 50 students. What do you think is the probability that at least two of them have the same birthday (ignoring the leap year complication)?

$$p = P(\text{at least one common birthday among the 50 students})$$
$$= 1 - P(\text{no common birthdays among the 50 students})$$
$$= 1 - (1)\left(\frac{364}{365}\right)\left(\frac{363}{365}\right)\cdots\left(\frac{316}{365}\right) = 1 - .030 = .970.$$

Wow–surprised?? Does this agree with your intuition? For class sizes 15, 18, 23, 40, and 70, the answers are $p = .253, .347, .507, .891,$ and $.999$, respectively. Let's try it with our class!

Note We have already seen that two events A and B are independent if and only if $P(A \cap B) = P(A)P(B)$. In conjunction with the Multiplication Rule in Lemma 2.3, this means that A and B are independent if and only if

$$P(A)\,P(B) = P(A \cap B) = P(A|B)\,P(B) = P(B|A)\,P(A),$$

which, in turn, holds if and only if

$$P(A) = P(A|B) \quad \text{and} \quad P(B) = P(B|A).$$

This, of course, totally agrees with our intuition about independent events and conditional probabilities. As we shall see later, however, we need to be cautious about our intuition around conditional probabilities!

Theorem 2.1 Theorem of Total Probability
Let $B_1, B_2, \ldots, B_k$ be a collection of k pairwise mutually exclusive and exhaustive events; that is,

$$B_i \cap B_j = \emptyset \text{ for } i \neq j \quad \text{and} \quad \bigcup_{i=1}^{k} B_i = S \text{ (sample space)}.$$

Then, for any other event A, we have

$$P(A) = \sum_{i=1}^{k} P(B_i)P(A|B_i).$$

Proof The events $A \cap B_1, \ldots, A \cap B_k$ are mutually exclusive and $A = \bigcup_{i=1}^{k} A \cap B_i$. It follows that

$$P(A) = \sum_{i=1}^{k} P(A \cap B_i) = \sum_{i=1}^{k} P(B_i)P(A|B_i). \quad \blacksquare$$

Example 2.21 Drawing Marbles
Consider the following three bowls of red and white marbles:

   Bowl 1: 4 red and 6 white
   Bowl 2: 8 white and 1 red
   Bowl 3: 8 red and 1 white

We draw a marble at random from Bowl 1 and place it in Bowl 2. Then we draw a marble at random from Bowl 2 and place it in Bowl 3. Finally, we draw a marble at random from Bowl 3. What is the probability that the marble drawn from Bowl 3 is red? Let

$$A = \{\text{marble drawn from Bowl 3 is red}\}$$

and

$$B_j = \{\text{marble drawn from Bowl } j \text{ is red}\}, j = 1, 2.$$

Then we have

$$A = (A \cap B_1 \cap B_2) \cup \left(A \cap B_1 \cap B_2^c\right) \cup \left(A \cap B_1^c \cap B_2\right) \cup \left(A \cap B_1^c \cap B_2^c\right),$$

so that A is a union of mutually exclusive and exhaustive sets. It follows from the Theorem of Total Probability that

$$P(A) = P(A \cap B_1 \cap B_2) + P(A \cap B_1 \cap B_2^c) + P(A \cap B_1^c \cap B_2) + P(A \cap B_1^c \cap B_2^c)$$
$$= P(\text{RRR}) + P(\text{RWR}) + P(\text{WRR}) + P(\text{WWR})$$
$$= \left(\frac{4}{10}\right)\left(\frac{2}{10}\right)\left(\frac{9}{10}\right) + \left(\frac{4}{10}\right)\left(\frac{8}{10}\right)\left(\frac{8}{10}\right) + \left(\frac{6}{10}\right)\left(\frac{1}{10}\right)\left(\frac{9}{10}\right) + \left(\frac{6}{10}\right)\left(\frac{9}{10}\right)\left(\frac{8}{10}\right)$$
$$= \frac{72 + 256 + 54 + 432}{1000} = .814.$$

Example 2.22 Will Our Battery Be Defective?
In a factory that produces AAA batteries, machines 1, 2, and 3 produce, respectively, 20%, 30%, and 50% of the total output for the factory. We also know that 5%, 3%, and 2% of the batteries produced by machines 1, 2, and 3, respectively, are defective. What is the probability that a battery purchased at random from this factory is defective?

$$P(\text{battery is defective})$$
$$= \sum_{i=1}^{3} P(\text{battery from machine } i)P(\text{defective}|\text{it came from machine } i)$$
$$= (.20)(.05) + (.30)(.03) + (.50)(.02) = .029.$$

Example 2.23 A Tale of Three Coins
In our pocket we have three coins, one of which has a head on both sides, while the other two are normal (fair) head/tail coins. A coin is chosen at random (without our knowledge of which coin it is) from the pocket and tossed three times. What is the probability that all three tosses result in heads?

$$P(\text{all three heads}) = P(\text{all three heads}|\text{two-headed coin}) \, P(\text{two-headed coin})$$
$$+ P(\text{all three heads}|\text{normal coin}) \, P(\text{normal coin})$$
$$= (1)(1/3) + [(1/2)(1/2)(1/2)] \, (2/3) = 5/12.$$

Theorem 2.2 Bayes Rule
Let $B_1, \ldots, B_k$ be a collection of k mutually exclusive and exhaustive events and let A be any other event. Then

$$P(B_j|A) = \frac{P(A \cap B_j)}{P(A)} = \frac{P(A|B_j)P(B_j)}{\sum_{i=1}^{k} P(A|B_i)P(B_i)}, \quad j = 1, \ldots, k.$$

Proof Let $j \in \{1, \ldots, k\}$ be arbitrary. Then, by the definition of conditional probability, we have

$$P(B_j|A) = \frac{P(A \cap B_j)}{P(A)}.$$

From the Theorem of Total Probability, we have

$$\frac{P(A \cap B_j)}{P(A)} = \frac{P(A|B_j)P(B_j)}{\sum\limits_{i=1}^{k} P(A|B_i)P(B_i)}$$

and the conclusion follows. ■

Example 2.24 Which Machine Is Guilty?

Consider the battery manufacturer discussed previously in Example 2.22. Suppose we purchase a single battery from this manufacturer and find it to be defective. What is the probability that it was produced by machine 1?

$$P(\text{battery from machine 1}|\text{defective}) = \frac{P(\text{battery is defective and from machine 1})}{P(\text{battery is defective})}$$

$$= \frac{P(\text{defective}|\text{machine 1}) \, P(\text{machine 1})}{P(\text{battery is defective})}$$

$$= \frac{10}{29} = .345$$

$$> .20 = P(\text{battery is from machine 1}).$$

Example 2.25 Was It the Two-Headed Coin?

Consider the three-coins-in-a-pocket setting discussed previously in Example 2.23. Suppose we independently flip the drawn coin three times and obtain all three heads. What is the probability that the two-headed coin was drawn and used for these three flips?

$P(\text{drawn coin is two-headed}|\text{a head on each of the three flips})$

$$= \frac{P \, (\text{two-headed coin and three heads})}{P(\text{three heads})} = \frac{1\left(\frac{1}{3}\right)}{5/12}$$

$$= \frac{4}{5} > P(\text{drawn coin is two-headed}) = \frac{1}{3}.$$

Example 2.26 Red Marbles on Draw 1 and Draw 3

Consider the three bowls of red and white marbles previously discussed in Example 2.21 and use the same process of successively drawing marbles. If the marble drawn from Bowl 3 is red, what is the probability that the marble drawn from Bowl 1 was also red?

P(marble drawn from Bowl 1 is red|marble drawn from Bowl 3 is red)

$$= \frac{P(\text{marbles drawn from Bowls 1 and 3 are both red})}{P(\text{marble drawn from Bowl 3 is red})} = \frac{P(\text{RWR or RRR})}{.814}$$

$$= \frac{[.072 + .256]}{.814} = \frac{.328}{.814} = .403 > .4 = P(\text{marble drawn from Bowl 1 is red}).$$

Example 2.27 Do Michigan Football Fans Drool?

Football games between THE Ohio State University and THE University of Michigan are among the oldest and most intense rivalries in the sport. Based on some recent surveys, the following are "well-known facts" (although they could be "alternative facts", a seemingly current fad):

> 70% of all Michigan football fans drool.
> 10% of all Ohio State football fans drool.
> 90% of the people in Columbus, Ohio, are Ohio State football fans.
> 10% of the people in Columbus, Ohio, are Michigan football fans.

If we choose a person at random from Columbus, Ohio, on the Friday night before one of these games and find that he/she drools, what is the probability that he/she is also a Michigan football fan? Using the "well-known facts", we have

$$P(\text{Michigan fan}|\text{drool}) = \frac{P(\text{Michigan fan and drool})}{P(\text{drool})} = \frac{(.10)(.70)}{(.10)(.70) + (.90)(.10)}$$

$$= \frac{.07}{.07 + .09} = \frac{7}{16} = .4375 < .5.$$

Thus, just because our randomly selected individual from Columbus, Ohio, drools is not sufficient cause to treat them as a dastardly Michigan football fan!

When we are dealing with conditional probabilities and, more specifically, with the application of Bayes rule, we must be very careful to understand exactly what is being computed and what is not being computed. It is not unusual for us to confuse what is being conditioned on, and this can lead to misleading inferences from carelessly inverting conditional probabilities in our minds.

Example 2.28 Is the Football Player Guilty?

Suppose we "know" the following information about campus restaurants/bars on a Saturday night following a football game:

(a) [Proportion of males at campus restaurants/bars on a Saturday night following a football game who are also football players] $= .0005$
(b) [Proportion of male football players at campus restaurants/bars on a Saturday night following a football game who get into a fight] $= .01$
(c) [Proportion of male non-football players at campus restaurants/bars on a Saturday night following a football game who get into a fight] $= .0005$

Note (b) is 20 times (c)

Now, suppose it turns out that a male chosen at random from the males at campus restaurants/bars on a Saturday night following a football game is involved in a fight. What is the probability that he is a FB player?

$$P(\text{FB}|\text{fight}) = \frac{P(\text{FB and fight})}{P(\text{fight})}$$

$$= \frac{P(\text{fight}|\text{FB})P(\text{FB})}{P(\text{fight}|\text{FB})P(\text{FB}) + P(\text{fight}|\text{not FB})P(\text{not FB})}$$

$$= \frac{(.01)(.0005)}{(.01)(.0005) + (.0005)(.9995)} = \frac{.000005}{.00050475} = .009906.$$

Thus, even though football players are 20 times more likely than non-football players to be involved in fights at campus restaurants/bars on Saturday evenings following a football game, if a randomly chosen male was actually involved in a fight at a campus restaurant/bar on a Saturday evening following a football game, **it is 99 times more likely that he is not a FB player**. Thus, the next time you hear about a fight at a campus restaurant/bar on a Saturday evening following a football game, DON'T immediately jump to the conclusion that a football player was involved in it.

Think About It What two totally different conditional probabilities lead to the possible confusion described in Example 2.28?

2.4   Exercises

2.1. Let S be a sample space for a random experiment and let A and B be two events associated with the experiment.

(a) Suppose A and B are disjoint events. Construct a Venn diagram that illustrates this disjointedness and then create additional Venn diagrams that clearly indicate each of the following regions:

$A \cap B, A^c, B^c, A \cup B, A \cap B^c, B \cap A^c, (A \cup B)^c, A^c \cup B^c, (A^c \cup B^c)^c$, and $A^c \cap B^c$.

(b) Suppose A and B are not disjoint events. Construct a Venn diagram that illustrates this fact and then create additional Venn diagrams that clearly indicate each of the following regions:

$A \cap B, A^c, B^c, A \cup B, A \cap B^c, B \cap A^c, (A \cup B)^c, A^c \cup B^c, (A^c \cup B^c)^c$, and $A^c \cap B^c$.

2.2. A particular city has two newspapers—one is delivered in the morning and one in the afternoon. Suppose that 75% of the households subscribe to the morning paper, 50% of the households subscribe to the afternoon paper, and 90% of the households subscribe to at least one of the papers.

(a) Construct a Venn diagram depicting this information.
(b) Show the following events on the Venn diagram constructed in part (a) (or create separate diagrams if that is preferable) and deduce the percentage of households belonging to each of the events:

1. {households that take both newspapers}
2. {households that take the morning newspaper but not the afternoon paper}
3. {households that take only the afternoon paper}
4. {households that take neither paper}

2.3. Construct an example where $P(B|A) > P(B)$.

2.4. Construct an example where $P(B|A) < P(B)$.

2.5. Construct an example where $P(B|A) = P(B)$.

2.6. We say that an event A contains negative or positive information about an event B if $P(B|A) < P(B)$ or $> P(B)$, respectively.

(a) Show that if an event A contains negative information about an event B, then the event B also contains negative information about the event A.
(b) Show that if an event A contains negative information about an event B, then the event A contains positive information about the event B^c.
(c) Prove or provide a counterexample: If an event A contains negative information about an event B, then the event A^c also contains negative information about the event B^c.

2.7. A fair coin is tossed as many times as necessary to obtain one head. Then the sample space is $S = \{H, TH, TTH, TTTH, ...\}$. Consider the probability set function $P(\cdot)$ that assigns the probabilities 1/2, 1/4, 1/8,... to the corresponding elements of S.

(a) Let $C_1 = \{H, TH, TTH, TTTH, TTTTH\}$. What is $P(C_1)$?
(b) Let $C_2 = \{TTTTH, TTTTTH\}$. Find $P(C_2)$, $P(C_1 \cap C_2)$, and $P(C_1 \cup C_2)$.

2.8. Let A and B be arbitrary events. Let $C = \{$exactly one of A or B occurs$\}$. Show that

$$P(C) = P(A) + P(B) - 2P(A \cap B).$$

2.9. For events A and B, find formulas for the probabilities of the following events in terms of $P(A)$, $P(B)$, and $P(A \cap B)$:

(a) $P(A$ or B or both)
(b) $P($either A or B but not both)
(c) $P($at least one of A or $B)$
(d) $P($at most one of A or $B)$

2.10. Let S be a sample space and let A_1, A_2, A_3 be mutually exclusive events such that $A_1 \cup A_2 \cup A_3 = S$. Suppose that $P(A_1) = .25$ and $P(A_2) = .12$.

(a) Find $P(A_1 \cup A_2)$.

(b) Find $P(A_1^c)$.

(c) Find $P(A_1^c \cap A_2^c)$.

2.11. Let A and B be independent events. Show that:

(a) A and B^c are also independent events

(b) A^c and B are also independent events

(c) A^c and B^c are also independent events

2.12. Suppose a **large** bag of candy (such as M & M's) contains the following color distribution.

Color	Brown	Red	Yellow	Green	Orange
Proportion	0.3	0.2	0.2	0.2	0.1

Three pieces of candy are drawn randomly from this bag and we are interested in the color of each piece of candy.

(a) List the set of all possible outcomes in the sample space S.

(b) Are the outcomes equally likely? Explain.

(c) What is the probability that the first piece of candy is red and the second piece of candy is orange?

(d) What is the probability that all three pieces of candy are the same color?

(e) What is the probability that all three pieces of candy are different colors?

2.13. Consider drawing a hand of five cards without replacement between draws from a standard deck of 52 playing cards (no jokers).

(a) What is the sample space for the outcome of your draws? How many possible outcomes are in the sample space?

(b) Describe two events that are disjoint and complementary.

(c) Describe two events that are disjoint but not complementary.

(d) Describe two events that are neither disjoint nor complementary.

(e) What is the natural probability set function for the subsets of the sample space?

2.14. Repeat Exercise 2.13, but now with replacement between draws.

2.15. Consider a bowl containing 16 pieces of candy, 4 caramels, 7 mints, and 5 chocolates. Without looking, you select two pieces of candy (without replacement between the selections, of course).

(a) What is the sample space for your two selections?

(b) What is the appropriate probability set function for the subsets of the sample space?

(c) What is the probability that you will choose two different types of candy?

(d) What is the probability that you will choose two pieces of the same type of candy?

(c) What is the probability that at least one of the selected pieces of candy is chocolate?

2.16. Consider the same candy bowl as in Exercise 2.15. Suppose now that you eat the first piece of candy, noting what type it is, before you select the second piece of candy.

(a) What are the sample spaces for your second selection given the type of candy in your first selection, and what are the appropriate probability set functions for this second selection?
(b) What is the probability that the second selection is the same as your first selection? How does this compare with your answer to part (d) in Exercise 2.15?

2.17. During the winter in Columbus, Ohio, the probability that a cloudy day is followed by another cloudy day is 0.8 and the probability that a sunny day is followed by a cloudy day is 0.6.

(a) Following a cloudy day on January 12 in Columbus, what is the probability that the weather conditions for January 13, 14, 15, and 16 are cloudy, sunny, sunny, and cloudy, respectively?
(b) Following a cloudy day on January 12 in Columbus, what is the probability that there are exactly two cloudy says in the next four consecutive days (January 13, 14, 15, and 16)?
(c) Given that it was cloudy on January 12 in Columbus and that exactly two of the next 4 days were also cloudy, what is the probability that the weather conditions for January 13, 14, 15, and 16 (following the cloudy day on January 12) were cloudy, sunny, sunny, and cloudy, respectively?

2.18. Each of the three football players (denote them by A, B, and C) will attempt to kick a field goal from the 30-yard line. Assume that the kicks are independent and that

$$P(A \text{ makes the field goal}) = 0.7,$$
$$P(B \text{ makes the field goal}) = 0.5,$$
$$P(C \text{ makes the field goal}) = 0.6.$$

(a) Find P(exactly one player successfully makes his field goal).
(b) Find P(exactly two players successfully make their field goals).

2.19. You have probability .6 and your friend has probability .7 of making any given free throw. You alternate shooting free throws until the first one is made, with you shooting first.
(a) What is the probability that the first free throw is made on the fifth shot?
(b) What is the probability that you make the first free throw?
(c) Given that the first free throw is made on the seventh shot, what is the probability that you make it?

(d) Given that the first free throw is made on or before the seventh shot, what is the probability that you make it?

2.20. A football team plays a season of six games. They are a very good team with their starting quarterback, winning 90% of their games if the starting quarterback plays the entire game. But their backup quarterback is very inexperienced and they only win 40% of their games when the backup quarterback plays the entire game. Suppose that P(starting quarterback is available to play) = .8 for each of the six games. Assume independence between the six games and that if the starting quarterback is available at the beginning of the game, he plays the entire game. Otherwise the backup quarterback plays the entire game.

(a) Let U denote the number of games in which the starting quarterback plays. Find the probability distribution for U.
(b) Let V denote the number of games that the team wins. Find the probability distribution for V.
(c) Let W denote the first game in which the backup quarterback plays. Find the probability distribution for W.
(d) Let Q denote the first game that the team wins. Find the probability distribution for Q.
(e) Suppose the team wins all six games. What is the probability that the starting quarterback played in all six of them?

2.21. Box 1 contains 7 red balls and 3 white balls; Box 2 contains 6 red balls and 3 white balls; Box 3 contains 4 red balls and 5 white balls. Draw one ball at random from Box 1 and place it in Box 2. Then draw one ball at random from this new Box 2 and place it in Box 3. Then draw one ball at random from this new Box 3.

(a) Find P(ball drawn from Box 3 is red).
(b) Find P(at least one of the three drawn balls is red).
(c) Find P(ball drawn from Box 3 is red | exactly one of the three drawn balls is red).
(d) Find P(exactly one drawn ball is read | ball drawn from Box 3 is red).

2.22. A bag contains five blue balls and three red balls. Someone draws a ball, and then a second ball is drawn (without replacement between the draws), and, finally, a third ball is drawn (again without replacement between the draws). Compute the following probabilities.

(a) P(no red balls left in the bag after the third draw)
(b) P(one red ball left in the bag after the third draw)
(c) P(first red ball is obtained on the third draw)
(d) P(a red ball is obtained on the third draw)
(e) Given that the ball obtained on the third draw is red, what is the probability that the first drawn ball was also red?
(f) Given that exactly two of the drawn balls are red, what is the probability that the first drawn ball was red?
(g) Given that at least two of the drawn balls are red, what is the probability that all three of the drawn balls are red?

2.23. Answer all seven parts of Exercise 2.22 if the three balls are drawn with replacement between each of the draws.

2.24. Answer all seven parts of Exercise 2.22 if the first drawn ball is replaced before the second ball is drawn, but there is no replacement between the second and third draws.

2.25. Answer all seven parts of Exercise 2.22 if the first drawn ball is not replaced before the second ball is drawn, but there is replacement of the second drawn ball before the third ball is drawn.

2.26. Some Little League baseball managers tend to be a bit irrational when it comes to calling heads or tails for the coin flip to determine home team at the beginning of a game. In the previous four games, Manager A has seen the coin come up heads each time. He figures that luck is on his side so he chooses heads on the next flip. Manager B, on the other hand, knows a bit of probability and he is aware that the probability is only $\left(\frac{1}{2}\right)^5 = \frac{1}{32}$ that a flipped fair coin will come up heads five times in a row, so he naturally chooses tails for the next flip. What would you like to tell these two managers?

2.27. Most homes have at least two smoke detectors. Suppose that the probability each smoke detector will function properly in the presence of smoke is .85 and that the smoke detectors function independently of one another.

(a) If you have two smoke detectors in your home, what is the probability that both of them function properly during a fire?
(b) If you have three smoke detectors in your home, what is the probability that exactly one of them will function properly during a fire?
(c) If you have two smoke detectors in your home, what is the probability that at least one of them will NOT function properly during a fire?
(d) Repeat part (c) if you have three smoke detectors in your home and compare the result with that from part (c).
(e) Suppose you have n (a positive integer) smoke detectors in your home. What is the probability that at least one of the smoke detectors will NOT function properly during a fire?
(f) Consider again the setting with n (a positive integer) smoke detectors in your home. Most importantly, what is the probability that at least one of the smoke detectors WILL function properly during a fire?

2.28. A survey organization asked American respondents their views on the likely future direction of the economy and whether they had voted for the current President in the last election. The two-way table below shows the proportion of responses in each category.

		View of economy		
		Optimistic	Pessimistic	Neutral
Voting behavior	For president	0.2	0.1	0.1
	Against president	0.1	0.15	0.05
	Did not vote	0.05	0.1	0.15

What is the probability that a randomly selected respondent:

(a) Voted against the President?
(b) Is pessimistic about the future of the economy?
(c) Voted for the President and is pessimistic about the future of the economy?
(d) Voted for the President but is not pessimistic about the future of the economy?
(e) Are the respondents' views on the economy and voting behavior independent? Explain.

2.29. A pair of fair dice are rolled one time and the sum of the numbers obtained is six. What is the probability that at least one of the dice came up a three?

2.30. A five-card hand is dealt from an ordinary 52-card deck (no jokers) and one card is turned up for all to see. If the turned-up card is a king, what is the probability that there is at least one more king among the remaining four cards in the hand?

2.31. Roll a pair of fair dice.

(a) Find P(doubles occur).
(b) Repeat the roll of the pair of fair dice n times. What is the probability that doubles occurs at least once in the n rolls?

2.32. Consider a draft lottery containing the 366 days of the year (including February 29).

(a) What is the probability that the first 180 days drawn (without replacement between draws) are evenly distributed across the 12 months?
(b) What is the probability that the first 30 days drawn contain no dates from September?

2.33. National Public Radio (2014) reported on a number of results from a survey conducted by the National Science Foundation in the United States in 2012. One of the questions asked in the survey was "Does the earth revolve around the sun, or does the sun revolve around the earth?" Twenty percent of the respondents said the sun revolved around the earth!! Suppose you randomly select ten individuals and ask them this same question. If the results of the National Science Foundation survey are applicable:

(a) What is the probability that at least one of the individuals you interview will believe that the sun revolves around the earth?
(b) What is the probability that exactly half of the individuals you interview will believe that the sun revolves around the earth?
(c) What is the probability that more than half of the individuals you interview will believe that the sun revolves around the earth?

2.34. At the beginning of many Little League tournament baseball games, the two opposing managers get together with the head umpire and flip a coin to determine which team will be home team. Manager A decides that this process is too simple. He makes the following proposition to the opposing Manager B (and the head umpire). First roll a fair six-sided die—suppose it results in an outcome $x \in \{1, 2, \ldots, 6\}$. Then the head umpire will independently flip a fair coin x times. Prior to the roll of the die, however, Manager A allows Manager B to choose either head or tail as his side for all x coin flips and states that Team A would be home team *only if* Manager A gets at least as many coin flips showing his side as Manager B gets for his choice. Should Manager B agree to the more interesting method for determining home team? Justify your answer.

2.35. Suppose a family has four children. Presuming independence between births and equal likelihood of a girl or boy on each birth, answer the following questions.

(a) What is the probability that all four children will be of the same sex?
(b) What is the probability that they will have exactly one girl?
(c) What is the probability that they will have exactly two boys and two girls?
(d) What is the probability that they will have at least one boy?
(e) What is the probability that they will have four boys, given that their first child is a boy?
(f) What is the probability that they will have four girls, given that at least one of them is a girl?
(g) What is the probability that they will have exactly two boys and two girls, given that at least one of them is a girl?
(h) What is the probability that they will have exactly three girls and one boy, given that at least one of them is a boy?

2.36. You have two coins in your pocket, one with a head on one side and a tail on the other side and one with a head on both sides. You randomly select one coin from your pocket and flip it once. Assume that each side is equally likely for both of the coins. If it comes up heads, you roll a fair six-sided die ten times. If it comes up tails, you roll the fair six-sided die only five times.

(a) What is the probability of no even numbers on the die rolls if you select the two-headed coin?
(b) What is the probability of no even numbers on the die rolls if you select the coin with a head on one side and a tail on the other side?
(c) What is the probability of no even numbers on the die rolls?
(d) If you observe no even numbers on your die rolls, what is the probability that you had selected the two-headed coin to flip?

2.37. If Emma and Ian study together, the probability is .9 that Ian will get an A on his Chemistry test. If Ted and Ian study together, however, the probability is only .5 that Ian will get an A on his Chemistry test—and if Ian cannot study with either of them and has to study alone, the probability drops to .3 that he will get an A on the Chemistry test. Unfortunately, all three students are working while going to school and their schedules are such that the probability is only .3 that Ian will be able to study with Emma, .6 that Ian will be able to study with Ted, and .1 that Ian will have to study alone (assume there is no chance that Ian can study with both Emma and Ted).

(a) What is the probability that Ian will get an A on his Chemistry test?
(b) If Ian gets an A on his Chemistry test, what is the probability that he had been able to find time to study with Emma? with Ted? had to study alone?

2.38. Flip a fair coin three independent times and let X be the number of heads in the three flips.

(a) What is the probability distribution for X?

Now flip the same coin three more independent times and let Y be the number of heads in the second three flips.

(b) Find $P(X = Y)$.

Now repeat this process of independently flipping three fair coins a total of n times and let X_i denote the number of heads on the ith set of three flips, $i = 1, 2, ..., n$.

(c) Find $P(X_1 = X_2 = \cdots = X_n)$ and obtain $\lim_{n \to \infty} P(X_1 = X_2 = \cdots = X_n)$.

2.39. Repeat Exercise 2.38 when the repeated event corresponds to flipping:

(a) Two fair coins
(b) Four fair coins

2.40. Suppose a family has ten children (all single births). Assume $P(\text{girl on any birth}) = \frac{1}{2}$.

(a) What is the probability that they have at least two girls?
(b) What is the probability that they have exactly k girls, for $k = 2, ..., 10$, given that they have at least two girls?
(c) What is the probability that they have at least q girls, for $q \in \{1, ..., 10\}$?
(d) What is the probability that they have exactly q girls, given that they have at least q girls, for $q = 1, ..., 10$?
(e) What is the probability that they have at least m girls, given that they have at least $q \in \{1, ..., 10\}$ girls, for $m = q, ..., 10$?

2.41. Two players, A and B, are competing at a quiz game involving a series of questions. For any individual question, the probabilities that A and B correctly answer the question are p and q, respectively, and the outcomes for different questions are assumed to be independent. The game ends when a player wins by answering a question correctly. Compute the probability that A wins the game if:

(a) Player A is given the first question in the game
(b) Player B is given the first question in the game

2.42. Two players, A and B, alternately and independently flip a coin, and the first player to obtain a head wins the game. Assume that Player A flips first.

(a) What is the probability that A wins if the coin is fair?
(b) Suppose that $P(\text{head}) = p$, not necessarily $\frac{1}{2}$. Find an expression (function of p) for the probability that A wins.
(c) Show that $P(A \text{ wins}) > 1/2$, for all $0 < p < 1$.

2.43. Consider a diagnostic test for the presence of a given disease. Suppose that the probability is quite high that the diagnostic test will correctly detect the disease if it is present in the patient. Suppose a friend of yours has this diagnostic test and it indicates that she has the disease. Should she immediately begin treatment for the disease without further testing? Describe the two conditional probabilities of interest here and discuss how the answer to this question depends on whether or not the disease is relatively common or rare.

2.44. It is a known fact that a large proportion of the Nazis in World War II were German. Suppose we were able to randomly choose a German from that period of time. Does this fact imply that we should immediately assume that the chosen German was a Nazi? Why or why not?

2.45. In the 1960s and 1970s (the "Hippie" era), it is a known fact that a large proportion of males who were using illicit drugs had long hair. (This is, of course, not true today!) Suppose we were able to randomly choose a male from the 1960s and 1970s who had long hair. Does this fact imply that we should immediately accuse the individual of using illicit drugs? Why or why not?

2.46. Suppose that 5% of men and .25% of women are color-blind. A person is chosen at random and that person is color-blind. What is the probability that the person is male? (Assume there are an equal number of males and females in the population.)

2.47. Suppose the probability is .85 that a lie detector test correctly identifies someone who is lying and that the probability is .10 that it incorrectly indicates that a person is lying when in fact they are telling the truth. If 95% of the individuals subjected to a lie detector test tell the truth, what is the probability an individual is telling the truth even though the lie detector indicates that he is lying?

2.48. Repeat Exercise 2.47 if only 50% of the individuals subjected to a lie detector test tell the truth. Comment on the difference between this answer and the one found in Exercise 2.47.

2.49. Consider a routine screening for a disease with a frequency of 0.5% in the population of interest. The accuracy of the test is quite high, with a false-positive rate (indicating presence of the disease when it is, in fact, not present) of only 4% and a false-negative rate (failing to detect the disease when it is, in fact, present) of only 8%. You are administered the test and the result comes back positive for the presence of the disease. Should you be concerned? What is the probability that you actually have the disease?

2.50. Consider the same setting described in Example 2.27, but suppose now that there is instead a 50–50 split among the residents of Columbus, Ohio (say it isn't so, Brutus!) in support of Michigan. Under this revised assumption, once again compute P(Michigan fan|drool). What accounts for the big difference here?

Chapter 3
Random Variables and Probability Distributions

In Chap. 2, we introduced the concept of probability and explored some of the basic properties of probability. In particular, we discussed how it pertains to specific simple events or combinations of simple events associated with the sample space of a random process. In most settings, however, we are interested in much more than just these specific events. In fact, it is important that we be able to obtain probabilities for all of the subsets of the sample space for a random process. This leads us to the concept of a **random variable** to describe the possible outcomes (i.e., sample space) for a random experiment and then naturally to the distribution of the associated probability across the subsets of this sample space (i.e., probability distribution) as described by possible values of the random variable.

> **Definition 3.1** Consider a random experiment whose outcome is a real number. Let X represent this random outcome and S the corresponding sample space for X. Then X is called a **random variable,** and the set of possible outcomes $x \in S$ is called the **support** or **domain of positive probability** for X.

In this chapter, we consider two distinctly different types of random variables.

Case 1 If the support for X is finite or at most countably infinite, it is called a **discrete random variable**.

Case 2 If the support for X is an interval or union of intervals, it is called a **continuous random variable**.

© Springer Nature Switzerland AG 2020
D. Wolfe, G. Schneider, *Primer for Data Analytics and Graduate Study in Statistics*,
https://doi.org/10.1007/978-3-030-47479-9_3

Note Sometimes various combinations of discrete and continuous variables can be of interest as well, but we will consider only these two separate cases in this text.

3.1   Discrete Random Variables

Definition 3.2 The **probability function for a discrete random variable X** with support S is a function $p_X(x)$ that satisfies the conditions

$$P(X = x) = p_X(x) \geq 0 \text{ for all } x \in S$$
$$= 0 \text{ for all } x \notin S$$

and

$$\sum_{x \in S} p_X(x) = 1.$$

We call

$$P(X = x), x \in S = p_X(x) \, I_S(x)$$

the **probability distribution** for X, where $I_S(x)$ is the indicator function for the sample space S.

It follows from the additive property of probabilities over disjoint unions that

$$P(X \in A) = \sum_{x \in A} p_X(x) \text{ for any subset } A \subset S.$$

Example 3.1 Rolling a Pair of Fair Dice
Let X be the sum of the numbers obtained on two rolls of a fair die. Then the sample space for X is

$$S = \{2, 3, \ldots, 11, 12\},$$

and the probability distribution for X is given by

$$P(X = x) = p_X(x) = \frac{1}{36}, x = 2, 12$$

$$= \frac{2}{36}, x = 3, 11$$

$$= \frac{3}{36}, x = 4, 10$$

$$= \frac{4}{36}, x = 5, 9$$

$$= \frac{5}{36}, x = 6, 8$$

$$= \frac{6}{36}, x = 7$$

$$= 0, \text{elsewhere}.$$

Thus, for example, it follows that

$$P(X \text{ is an even number}) = \sum_{x \text{ even}} p_X(x) = \frac{1 + 1 + 3 + 3 + 5 + 5}{36} = \frac{1}{2}$$

and

$$P(X \le 3) = \sum_{x=2}^{3} p_X(x) = \frac{1 + 2}{36} = \frac{3}{36}.$$

At several junctures throughout this text, we will be relying on some basic mathematical properties of real numbers that are typically discussed in a finite mathematics and/or an advanced calculus course. At these points in the text, we will reintroduce you to the necessary mathematical results that will be used in the ensuing discussion through the concept of **Mathematical Moments**.

Mathematical Moment 1 Geometric Series

Let a and r be arbitrary numbers and consider the sequence of numbers

$$a, ar, ar^2, ar^3, \ldots \ldots$$

We call this a **geometric progression** with initial term a and multiplicative term r. Let n represent the number of terms in the geometric progression, t_n the nth term in the progression, and S_n the sum of the first n terms in the progression. Then we have

$$t_n = ar^{n-1} \quad \text{and} \quad S_n = a + ar + ar^2 + \ldots + ar^{n-1} = a \sum_{i=1}^{n} r^{i-1}.$$

But $(r - 1)(1 + r + \cdots + r^{n-1}) = (r + r^2 + \cdots + r^n - 1 - r - r^2 - \cdots - r^{n-1}) = r^n - 1$, which implies that

$$\left[1 + r + \cdots + r^{n-1}\right] = \frac{r^n - 1}{r - 1},$$

so that

$$S_n = a\frac{r^n - 1}{r - 1}.$$

Also, if $|r| < 1$, then we have $s_\infty = \lim_{n \to \infty} S_n = \lim_{n \to \infty} \left[a\frac{r^n - 1}{r - 1}\right] = -\frac{a}{r-1}$, so that

$$\sum_{i=1}^{\infty} ar^{i-1} = \frac{a}{1-r}.$$

Example 3.2 Bernoulli Trial and Geometric Random Variable

Definition 3.3 Consider an experiment that can result in outcome A with probability $p > 0$ and not A with probability $(1 - p)$. Such an experiment is called a **Bernoulli trial with probability of "success" (i.e., outcome A)** p. Suppose we conduct such Bernoulli experiments independently until the first outcome A occurs. Let the random variable X denote the trial on which this happens. Then X is called a **geometric random variable with parameter p,** and it has support space $S = \{1, 2, 3, \ldots\}$ and probability function

$$P(X = x) = p_X(x) = P(\text{not A}, \text{not A}, \ldots, \text{not A}, \text{A})$$
$$= p(1 - p)^{(x-1)} I_{\{1,2,3,\ldots\}}(x) \tag{3.1}$$

We use the notation $X \sim Geom(p)$ to denote that X has a geometric distribution with parameter p.

Now, suppose that $X \sim Geom(p)$. Then we have

$$P(X \text{ is an even number}) = \sum_{x=2(2)}^{\infty} p(1 - p)^{x-1} = p(1 - p) + p(1 - p)^3 + \cdots$$

$$= \frac{p(1 - p)}{1 - (1 - p)^2},$$

since this is just a geometric series with initial term $p(1 - p)$ and multiplicative term $(1 - p)^2$. Also,

$$P(X \text{ is an odd number}) = \sum_{x=1(2)}^{\infty} p(1 - p)^{x-1} = p + p(1 - p)^2 + \cdots = \frac{p}{1 - (1 - p)^2},$$

since this is just another geometric series with initial term p and multiplicative term $(1 - p)^2$. (Also, of course, $P(X$ is an odd number$) = 1 - P(X$ is an even number).)

In the same vein, we have

$$P(X < 10) = P(X \leq 9) = \sum_{x=1}^{9} p(1-p)^{x-1}$$

$$= S_9 = p\frac{(1-p)^9 - 1}{(1-p-1)} = 1 - (1-p)^9$$

$$= 1 - P(X \geq 10) = 1 - \sum_{x=10}^{\infty} p(1-p)^{x-1}$$

$$= 1 - \frac{p(1-p)^9}{1 - (1-p)} = 1 - (1-p)^9.\text{(Why?)}$$

Note that

$$\sum_{x=1}^{\infty} P(X = x) = \sum_{x=1}^{\infty} p(1-p)^{x-1} = \frac{p}{1 - (1-p)} = 1,$$

as it must be since this is just a geometric series with initial term p and multiplicative term (1-p).

Example 3.3 Tailgating

The AAA Foundation of Public Safety reported (July, 2016) that 50.8% of drivers acknowledged having aggressively tailgated another vehicle to express displeasure at least once during 2014. Consider a sequence of cars following you on a major highway. Assuming randomness of the individual drivers on the highway and independence between the drivers, what is the probability that the driver of the first car behind you had never previously aggressively tailgated, but that the driver of the second car behind you had, in fact, previously aggressively tailgated? The event of interest here is that the closest driver who had previously aggressively tailgated is in the second car following you. Given our assumption about the randomness and independence between individual drivers on the highway, the drivers in the cars following you on the highway represent a sequence of independent Bernoulli trials with probability of "success" (i.e., having previously aggressively tailgated!) $p = .508$ for each trial. Letting X denote the sequential number of the first driver following you who had previously aggressively tailgated, we have that $X \sim Geom$ (.508). It follows that

$$P(X = 2) = (.508)(1 - .508)^{2-1} = (.508)(.492) = .250.$$

Think About It What do you think about the reasonableness of the assumption of randomness and independence for the tailgating behavior of individual drivers behind you on the highway?

You can also use the **R** function *dgeom( )* to calculate the value of the probability distribution for a geometric random variable by specifying the arguments *x* and *prob*, which are the number of failures (not the number of trials!) and probability of success, respectively. Running the following command verifies the result we obtained in Example 3.3.

> dgeom($x = 1$, *prob* = 0.508)

[1] 0.249936

We can also compute the probability that the number of failures is less than or equal to a value *q* using the **R** function *pgeom( )*. For example, to compute $P(X \leq 5)$, we can use

> pgeom($q = 4$, *prob* = 0.508)

[1] 0.9711713,

recalling that "less than or equal to 5 trials" is equivalent to "less than or equal to 4 failures". For now, this function makes life more convenient by saving us the need to call the *dgeom( )* function multiple times and then adding up the results of these function calls. In later chapters, we will see that this function for calculating cumulative probability is also of interest in its own right.

Mathematical Moment 2 Taylor/Maclaurin Series

Definition 3.4 A **Taylor series** is a representation of a function as an infinite sum of terms that are calculated from the values of the function's derivatives at a single point. The concept of a Taylor series was formally introduced by the English mathematician Brook Taylor in 1715. If the Taylor series is centered at zero, then the series is also called a **Maclaurin series**, named after the Scottish mathematician Colin Maclaurin, who made extensive use of this special case of Taylor series in the 18th century. It is common practice to approximate a function by using a finite number of terms of its Taylor series.

Result Let $f(x)$ be a real valued function defined on $[c, d]$ such that $f(\cdot)$ and its first $(n+1)$ derivatives exist and are continuous at all $x \in [c, d]$. Let $a, x \in [c, d]$. Then there is a value ξ between a and x such that

$$f(x) = f(a) + \frac{f'(a)(x-a)}{1!} + \frac{f''(a)(x-a)^2}{2!} + \cdots + \frac{f^{(n)}(a)(x-a)^n}{n!} + \frac{f^{(n+1)}(\xi)(x-a)^{n+1}}{(n+1)!}.$$

This is called the **Taylor series expansion of $f(x)$ about the point $x = a$.** If the function $f(\cdot)$ has unlimited derivatives, then the complete Taylor series expansion for $f(x)$ about $x = a$ can be written as:

$$f(x) = f(a) + \frac{f'(a)(x-a)}{1!} + \frac{f''(a)(x-a)^2}{2!} + \cdots + \frac{f^{(n)}(a)(x-a)^n}{n!} + \cdots$$

$$= f(a) + \sum_{n=1}^{\infty} \frac{f^{(n)}(a)}{n!}(x-a)^n.$$

Special Case If we take $a = 0$, then the Taylor series expansion of $f(x)$ about $a = 0$ is given by

$$f(x) = f(0) + \frac{f'(0)x}{1!} + \frac{f''(0)x^2}{2!} + \cdots + \frac{f^{(n)}(0)x^n}{n!} + \cdots$$

$$= f(0) + \sum_{n=1}^{\infty} \frac{f^{(n)}(0)}{n!}x^n.$$

(3.2)

This is the Maclaurin series for a function $f(x)$ that has unlimited derivatives. Note that it is simply a polynomial power series in x.

Example 3.4 Taylor/Maclaurin Series for $f(x) = e^x$
Consider the function $f(x) = e^x$. Then we have $f^{(t)}(x) = e^x$, for all $t = 1, 2, \ldots$, and the Maclaurin series for e^x is given by

$$f(x) = e^x = e^0 + \frac{e^0 x}{1!} + \frac{e^0 x^2}{2!} + \cdots + \frac{e^0 x^n}{n!} + \cdots$$

$$= 1 + \frac{x}{1!} + \frac{x^2}{2!} + \cdots + \frac{x^n}{n!} + \cdots = \sum_{t=0}^{\infty} \frac{x^t}{t!}.$$

and this Maclaurin expansion is valid for all real x.

Example 3.5 Binomial Expansion
Let c and x be arbitrary numbers, and let n be a positive integer. Consider the function $f(x) = (c + x)^n$, for $-\infty < x < \infty$. Then, we have

$$f^{(1)}(x) = n(c+x)^{n-1}, f^{(2)}(x) = n(n-1)(c+x)^{n-2}, \ldots,$$

$$f^{(t)}(x) = n(n-1)\cdots(n-t+1)(c+x)^{n-t}$$

for $t = 1, \ldots, n$, and $f^{(t)}(x) = 0$, for all $t > n$. It follows that the Maclaurin series expansion of $f(x) = (c + x)^n$ is given by

$$f(x) = (c+x)^n = f(0) + \sum_{t=1}^{\infty} \frac{f^{(t)}(0)}{t!} x^t$$

$$= c^n + nc^{n-1}x + \frac{n(n-1)}{2!} c^{n-2}x^2 + \cdots + \frac{n!}{n!} c^{n-n}x^n$$

$$= \binom{n}{0} c^n x^0 + \binom{n}{1} c^{n-1}x + \binom{n}{2} c^{n-2}x^2 + \binom{n}{3} c^{n-3}x^3 + \cdots + \binom{n}{n-1} c^1 x^{n-1} + \binom{n}{n} c^0 x^n$$

$$= \sum_{t=0}^{n} \binom{n}{t} c^{n-t} x^t,$$

which is just the well-known binomial expansion of $(c + x)^n$.

3.2 Discrete Random Variables

Example 3.6 Binomial Distribution

Definition 3.5 Consider conducting n independent Bernoulli experiments with probability p of "success," and let X be the number of successes obtained in these n Bernoulli trials. X has support space $S = \{0, 1, 2, \ldots, n\}$, and it is called a **Binomial random variable with parameters n and p**. We denote this by $X \sim Binom\,(n, p)$. It has probability function given by

$$P(X = \mathrm{x}) = p_X(x) = \binom{n}{x} p^x (1 - p)^{n-x} I_{\{0,1,\ldots,n\}}(x). \qquad (3.3)$$

Thus, if $X \sim Binom\,(5, .3)$, it follows that

$$P(X \text{ is an even number}) = P(X = 0) + P(X = 2) + P(X = 4)$$

$$= \binom{5}{0}(.3)^0(.7)^5 + \binom{5}{2}(.3)^2(.7)^3 + \binom{5}{4}(.3)^4(.7)^1$$

$$= .16807 + .3087 + .02835 = .50512.$$

Note $\sum_{x=0}^{n} P(X = x) = \sum_{x=0}^{n} \binom{n}{x} p^x (1 - p)^{n-x} = [p + (1 - p)]^n = 1^n = 1$, since this is just the binomial expansion of $[p + (1 - p)]^n$.

Think About It Is this result a surprise?

Example 3.7 Landlines Versus Cell Phones
According to a survey by GfK (2015), 44% of adults in the United States live in households with cell phones but no landline. Suppose you survey ten randomly

chosen adults. What is the probability that at most four of these adults live in households with cell phones but no landline?

Let X be the number of adults you surveyed who live in households with cell phones but no landline. Then we know that $X \sim Binom$ (10, .44), and it follows that

$$P(X \leq 4) = \sum_{x=0}^{4} (.44)^x (.56)^{10-x}$$

$$= \left[(.56)^{10} + \binom{10}{1}(.44)^1(.56)^9 + \binom{10}{2}(.44)^2(.56)^8 + \binom{10}{3}(.44)^3(.56)^7 \right.$$

$$\left. + \binom{10}{4}(.44)^4(.56)^6 \right]$$

$$= .0030 + .0238 + .0843 + .1765 + .2427 = .5303.$$

You can easily compute this quantity for yourself by using the **R** function *pbinom( )* and specifying the *q*, *size*, and *prob* arguments, which represent the number of successes, number of trials, and probability of success, respectively. Running the following command verifies the result we obtained in Example 3.7.

```
> pbinom(q = 4, size = 10, prob = 0.44)
[1] 0.5304187
```

Note that we can also compute the probability that X is **exactly** equal to a given value of x using the **R** function *dbinom( )*. For example, to compute $P(X = 4)$, we can use the command

```
> dbinom(x = 4, size = 10, prob = 0.44)
[1] 0.2427494
```

Example 3.8 Poisson Distribution

Definition 3.6 Let $\lambda > 0$ be a constant, and let X be a discrete random variable with support $S = $ nonnegative integers $= \{0, 1, 2, ...\}$ and probability function given by

$$P(X = x) = \frac{\lambda^x}{x!} e^{-\lambda} I_{\{0,1,2,...\}}(x). \tag{3.4}$$

We say that such an X is a **Poisson random variable with parameter** λ, and we write $X \sim Poisson$ (λ).

Thus, if $X \sim Poisson$ (2), it follows that

$$P(X \leq 3) = \sum_{x=0}^{3} \frac{2^x}{x!} e^{-2} = e^{-2}(1 + \frac{2}{1!} + \frac{2^2}{2!} + \frac{2^3}{3!}) = e^{-2}(1 + 2 + 2 + \frac{4}{3}) = \frac{19e^{-2}}{3}.$$

Think About It Let $X \sim Poisson$ (λ). Since the support for X is $\{0, 1, 2, ...\}$, it must be the case that $\sum_{x=0}^{\infty} P(X = x) = \sum_{x=0}^{\infty} \frac{\lambda^x}{x!} e^{-\lambda} = 1$. How do we know that this is true?

You can easily compute this quantity for yourself by using the **R** function *ppois( )* and specifying the q and *lambda* arguments, which represent the counts and lambda parameter, respectively. Running the following command verifies the result we obtained in Example 3.8.

```
> ppois(q = 3, lambda = 2)
[1] 0.8571235
```

Note that we can also compute the probability that X is **exactly** equal to a given value of x using the **R** function *dpois( )*. For example, to compute $P(X = 1)$, we can use the command

```
> dpois(x = 1, lambda = 2)
[1] 0.2706706
```

Example 3.9 Negative Binomial Distribution

Definition 3.7 Consider conducting independent Bernoulli experiments with common probability p of "success" until the mth ($m \geq 1$) success occurs. Let the random variable Y denote the trial on which this happens. Then Y is called a **negative binomial random variable with parameters m and p**. We write this as $Y \sim NegBin$ (m, p). It has support space $S = \{m, m+1, m+2, ...\}$ and probability function

$$P(Y=y)=p_Y(y)$$
$$=P(m-1 \text{ successes in the first } y-1 \text{ trials and a success on the } y\text{th trial})$$
$$\overset{ind}{=} P(m-1 \text{ successes in the first } y-1 \text{ trials})P(\text{success on the } y\text{th trial})$$
$$= \binom{y-1}{m-1}p^{m-1}(1-p)^{y-m} \times p\, I_{\{m,m+1,m+2,...\}}(y)(\text{Why?})$$
$$= \binom{y-1}{m-1}(1-p)^{y-m}p^m I_{\{m,m+1,m+2,...\}}(y)$$

$$(3.5)$$

Example 3.10 Landlines Versus Cell Phones Revisited

Consider the same setting as for Example 3.7. What is the probability that you survey no more than four adults before you reach the third person who does **not** live in a household with cell phones but no landline? Let Y be the number of adults surveyed before you reach the third person who does not live in a household with cell phones but no landline. Then Y has a negative binomial distribution with $m = 3$ and $p = .56$, since here a success occurs when you survey an adult who does not live in a household with cell phones but no landline. Thus, the probability of interest is

$$P(Y \le 4) = \sum_{y=3}^{4} \binom{y-1}{2} (.44)^{y-3}(.56)^3$$

$$= (.56)^3 + \binom{3}{2} (.44)^1(.56)^3$$

$$= (.56)^3[1 + 3(.44)] = .4074.$$

This computation is not all that difficult, but suppose, instead, that we wanted to know the probability that we would have to survey at least ten adults before we reached the third person who does not live in a household with cell phones but no landline. The expression for this probability is easy enough to write down:

$$P(Y \ge 10) = 1 - P(Y \le 9) = 1 - \sum_{y=3}^{9} \binom{y-1}{2} (.44)^{y-3}(.56)^3$$

However, it is a bit cumbersome to obtain a numerical value for this probability. Once again, software comes to the rescue, as you can easily compute this quantity for yourself by using the **R** function *pnbinom( )* and specifying the *q, size,* and *prob* arguments, representing the number of failures, the number of successes, and the probability of success, respectively. Running the following command verifies the result we obtained in Example 3.10.

```
> pnbinom(q = 1, size = 3, prob = 0.56)
[1] 0.4074291
```

We can also compute the probability that Y is **exactly** equal to a given value using the **R** function *dnbinom( )* in a manner similar to what we have seen for other **R** functions.

Finally, to compute $P(Y \ge 10)$, we can once again call *pnbinom( )*, but now with the *lower.tail* argument specified to be *FALSE*, which then gives $P(Y > y)$. Note that for this example, we specify q to be 6, since **R** gives $P(Y > q + size) = P(Y > 9) = P(Y \ge 10)$.

> pnbinom($q = 6$, $size = 3$, $prob = 0.56$, $lower.tail$ = FALSE)

[1] 0.04374361

Notes

3.1. The geometric probability distribution with parameter p is simply a special case of the negative binomial distribution with parameters $m = 1$ and p.

3.2. The binomial and negative binomial probability distributions are both based on independent Bernoulli trials with common probability of success p. The difference is that the number of trials is fixed and the number of successes is random for the binomial distribution, while the number of successes is fixed and the number of trials is random for the negative binomial distribution. This will be an important distinction to keep in mind when you are describing and solving associated problems with Bernoulli trials.

Example 3.11 Hypergeometric Distribution

Definition 3.8 Consider a population of N items, b of one type (say, "good") and $N - b$ of a second type (say, "not good"). We draw a sample of n items at random from the population **without replacement between the draws,** and let X be the number of "good" items in our sample. Then the support for X is given by $S = $ maximum $(0, n - N + b)$, ..., minimum (b, n). We call X a **hypergeometric random variable with parameters N, b, and n,** and it has probability function

$$P(X = x) = p_X(x) = \frac{\binom{b}{x}\binom{N-b}{n-x}}{\binom{N}{n}}I_S(x) \qquad (3.6)$$

Think About It Why is $S = $ maximum $(0, n - N + b)$, ..., minimum (b, n) the correct support for the hypergeometric random variable with parameters N, b, and n?

Example 3.12 Bridge Hands

Randomly deal a 13-card bridge hand from an ordinary deck of 52 cards (without the jokers). Let $X = $ number of spades in the hand. Then $N = 52$, $b = 13$, $n = 13$, and X has a hypergeometric distribution with these parameters and probability function

$$p_X(x) = \frac{\binom{13}{x}\binom{39}{13-x}}{\binom{52}{13}}I_{\{1,2,\dots,13\}}(x).$$

Note that here the support space $S = \{0, 1, \dots, 13\}$ is unconstrained, since $n = $ min (b, n) and $0 = $ max $(0, n - N + b)$.

One exceptional bridge hand occurs when all 13 of our cards are spades. However, the probability of this happening is virtually zero, since

$$P(X = 13) = \frac{\binom{13}{13}\binom{39}{0}}{\binom{52}{13}} = \frac{1}{\binom{52}{13}} = \frac{1}{635013559600} = 1.575\, e^{-12},$$

which might take you several lifetimes of constantly playing bridge to ever see in your hand!

Clearly, hypergeometric probabilities can be very difficult to compute numerically. However, for most hypergeometric distributions, we can use **R** to once again assist us. In particular, you can easily compute $P(X = 13)$ in Example 3.12 for yourself by using the **R** function *dhyper( )* and specifying the *x*, *m*, *n*, and *k* arguments, representing the number of successes, number of "good" items, number of "not good" items, and number of sample items drawn, respectively. (Be careful to note that these arguments for the **R** function *dhyper( )*—and later for the **R** function *phyper( )*—use different notation for the parameters of the hypergeometric distribution and the sample size than we have used in this text.) Running the following command as indicated verifies the result we obtained in Example 3.12.

```
> dhyper(x = 13, m = 13, n = 39, k = 13)
[1] 1.57477e-12
```

As we have seen before, where there is a *d-* function, there is generally a corresponding *p-* function. In this case, we can use the **R** function *phyper( )* to compute the probability of obtaining no more than three spades, namely, $P(X \le 3)$.

```
> phyper(q = 3, m = 13, n = 39, k = 13)
[1] 0.5850558
```

Example 3.13 Drawing Balls from an Urn

Consider an urn with eight red balls and two white balls. We draw three balls at random from the urn without replacement between the draws.

Case 1 Suppose we identify "good" with a white ball, so that the variable of interest here is $X =$ number of white balls drawn. In this case, we have $b = 2$, $n = 3$, and $N = 10$, so that X has a hypergeometric distribution with parameters $N = 10$, $b = 2$, and $n = 3$. Thus $n - N + b = 3 - 10 + 2 = -5 < 0$ so that maximum $(0, n - N + b) =$ maximum $(0, -5) = 0$, and the lower limit for S is not constrained. However, minimum $(b, n) =$ minimum $(2, 3) = 2 = b$, so that the upper limit for S is constrained to be 2, not 3, since there are only 2 white balls in the urn. Here the probability distribution for X is given by

$$p_X(x) = \frac{\binom{2}{x}\binom{8}{3-x}}{\binom{10}{3}} I_{\{0,1,2\}}(x).$$

Case 2 On the other hand, if we identify "good" with a red ball, then $X =$ number of red balls drawn. In this case, we have $b = 8$, $n = 3$, and $N = 10$, so that X has a hypergeometric distribution with parameters $N = 10$, $b = 8$, and $n = 3$. Here minimum $(n, b) =$ minimum $(3, 8) = n = 3$, and the upper limit for S is not constrained. However, $n - N + b = 3 - 10 + 8 = 1 > 0$ so that maximum $(0, n - N + b) =$ maximum $(0, 1) = 1$, and the lower limit for S is constrained to be 1, not 0. In this case, the probability distribution for X is given by

$$p_X(x) = \frac{\binom{8}{x}\binom{2}{3-x}}{\binom{10}{3}} I_{\{1,2,3\}}(x).$$

Note Both the hypergeometric and binomial probability distributions can be applied to a setting where we sample from a binary population with fixed numbers of "good" and "not good" items. The binomial distribution applies when we draw items from the population with replacement between draws, so that the proportion, p, of "good" items stays constant from draw to draw. The hypergeometric distribution applies when we draw items from the population without replacement between draws, so that the proportion, p, of "good" items changes from draw to draw. Notice also that while the draws are independent when we have replacement between draws (binomial), this is no longer the case when we do not have replacement between draws (hypergeometric).

3.3   Continuous Random Variables

Let X represent the outcome of an experiment, and let S denote the associated sample space. If S is an interval or union of intervals, we say that X is a **continuous random variable**.

Definition 3.9 The **probability density function (p.d.f.) for a continuous random variable X** with support S is a function $f_X(x)$ such that:

(a) $f_X(x) \geq 0$ for all $x \in S$

(b) $f_X(x) = 0$ for all $x \notin S$

(c) $P(X \in A) = \int_A f_X(x) \, dx$, for any $A \subset S$

(continued)

Definition 3.9 (continued)

and

$$(d)\ P(X \in S) = 1.$$

We call

$$f_X(x)\, I_S(x)$$

the **probability distribution** for X, where $I_S(x)$ is the indicator function for the sample space S.

Example 3.14 Uniform Distribution

Definition 3.10 Let X represent the random draw of a number from the interval $(a,\ b)$. Then the sample space (support) is $S = (a,\ b)$, and the p.d.f. for X is given by

$$f_X(x) = \frac{1}{b - a} I_{(a,b)}(x). \tag{3.7}$$

We say that X has a **uniform distribution over the interval (a, b),** and we write $X \sim Unif\,(a,\ b)$. For any $a < c < d < b$, it follows that

$$P(c < X < d) = \int_c^d \frac{1}{b - a}\,dx = \frac{d - c}{b - a}.$$

Example 3.15
Let X be a continuous random variable with support $S = (0,\ 1)$ and p.d.f.:

$$f_X(x) = 2x\, I_{(0,1)}(x).$$

Then,

$$P(X < \tfrac{1}{2}) = \int_0^{1/2} 2x\,dx = x^2\,\Big|_0^{1/2} = \frac{1}{4} - 0 = \frac{1}{4},$$

$$P(X > \tfrac{3}{4}) = \int_{3/4}^{1} 2x\,dx = x^2 \Big|_{3/4}^{1} = 1 - \frac{9}{16} = \frac{7}{16},$$

and

$$P(\tfrac{1}{2} < X < \tfrac{3}{4}) = \int_{1/2}^{3/4} 2x\,dx = x^2 \Big|_{1/2}^{3/4} = \frac{9}{16} - \frac{1}{4} = \frac{5}{16}.$$

Note also that $P\!\left(\tfrac{1}{2} < X < \tfrac{3}{4}\right) = 1 - P\!\left(X < \tfrac{1}{2}\right) - P\!\left(X > \tfrac{3}{4}\right) = 1 - \tfrac{1}{4} - \tfrac{7}{16} = \tfrac{5}{16}$.

Comment Let X be a continuous random variable with p.d.f. $f_X(x)$. Then, for any $A \subset S$,

$$P(X \in A) = \int_A f_X(x)\,dx = \text{area under the p.d.f.curve over the set A.}$$

Example 3.16

Consider the random variable in Example 3.15. Then $P(\tfrac{1}{2} < X < \tfrac{3}{4}) =$ area under the curve $f_X(x) = 2x\, I_{(0,\,1)}(x)$ between $x = \tfrac{1}{2}$ and $x = \tfrac{3}{4}$. We can see from Fig. 3.1 that the probability under the curve can be broken into two pieces—a rectangle with width 0.25 and height 1 and a triangle with base length 0.25 and height 0.5. Thus, we have $P(\tfrac{1}{2} < X < \tfrac{3}{4}) = 1 \times (0.25) + (0.5) \times (0.25) \times (0.5) = 0.3125$.

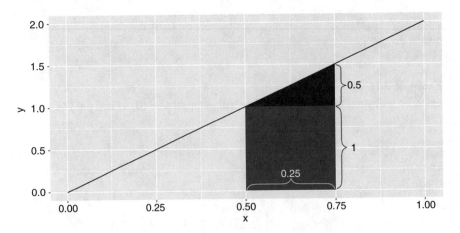

Fig 3.1 Calculating a probability for a continuous distribution by using geometry to find the area under the probability density function

Example 3.17 Gamma Distribution

Definition 3.11 Let X be a continuous random variable with p.d.f.:

$$f_X(x) = \frac{1}{\Gamma(\alpha)\beta^\alpha} x^{\alpha-1} e^{-\frac{x}{\beta}} I_{(0,\infty)}(x), \tag{3.8}$$

where $\alpha > 0$ and $\beta > 0$ are fixed and $\Gamma(t)$ is the gamma function defined by

$$\Gamma(t) = \int_0^\infty y^{t-1} e^{-y} dy, \text{ for } t > 0. \tag{3.9}$$

We say that such an X has a **Gamma distribution with parameters α and β**, and we write $X \sim Gamma\ (\alpha, \beta)$.

The **R** function *pgamma( )* can be used (with the *shape* and *scale* arguments corresponding to α and β, respectively) to calculate $P(X < b)$ and $P(X < a)$, the difference of which yields $P(a < X < b)$, for $X \sim Gamma\ (\alpha, \beta)$, as shown in Fig. 3.2.

Example 3.18 Chi-Square Distribution

Definition 3.12 If $X \sim Gamma\ \left(\alpha = \frac{r}{2}, \beta = 2\right)$ for some positive integer r, we say that X has a **chi-square distribution with r degrees of freedom**, and we write $X \sim \chi^2(r)$. It has p.d.f.:

$$f_X(x) = \frac{1}{\Gamma\left(\frac{r}{2}\right) 2^{\frac{r}{2}}} x^{\frac{r}{2}-1} e^{-\frac{x}{2}} I_{(0,\infty)}(x). \tag{3.10}$$

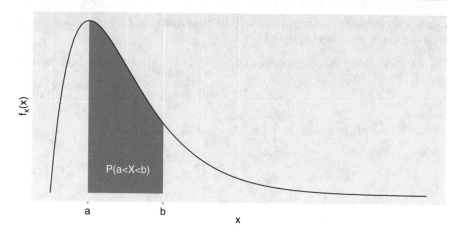

Fig 3.2 Calculating $P(a < X < b)$ for $X \sim Gamma\ (\alpha, \beta)$

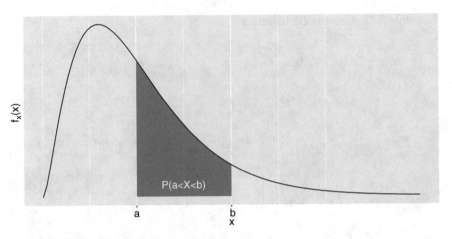

Fig 3.3 Calculating $P(a < X < b)$ for $X \sim \chi^2(r)$

The **R** function *pchisq( )* can be used (with the *df* argument corresponding to *r*) to calculate $P(X < b)$ and $P(X < a)$, the difference of which yields $P(a < X < b)$, for $X \sim \chi^2(r)$, as shown in Fig. 3.3. Note the similarity in shape between this chi-square density and the gamma density shown in Fig. 3.2.

Example 3.19 Exponential Distribution

Definition 3.13 If $X \sim Gamma\ (\alpha = 1, \beta)$, we say that **X has an exponential distribution with parameter $\beta > 0$**, and we write $X \sim Exp\ (\beta)$. It has p.d.f.:

$$f_X(x) = \frac{1}{\beta} e^{-\frac{x}{\beta}} I_{(0,\infty)}(x). \tag{3.11}$$

The **R** function *pexp( )* can be used (with the *rate* argument corresponding to $\frac{1}{\beta}$ – be careful!) to calculate $P(X < b)$ and $P(X < a)$, the difference of which yields $P(a < X < b)$, for $X \sim Exp\ (\beta)$, as shown in Fig. 3.4.

For the exponential distribution, we can explicitly calculate relevant probabilities directly. Thus, if $X \sim Exp\ (\beta)$, then

$$P(a < X < b) = \int_a^b \frac{1}{\beta} e^{-\frac{x}{\beta}} dx = -e^{-\frac{x}{\beta}} \Big|_{x=a}^{x=b} = e^{-\frac{a}{\beta}} - e^{-\frac{b}{\beta}}.$$

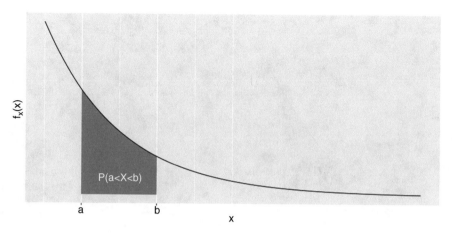

Fig 3.4 Calculating $P(a < X < b)$ for $X \sim \text{Exp}\,(\beta)$

Example 3.20 Normal (Gaussian) Distribution

Definition 3.14 Let X be a continuous random variable with p.d.f.:

$$f_X(x) = \frac{1}{\sqrt{2\pi\tau^2}}\, e^{-\frac{1}{2\tau^2}(x-\theta)^2} I_{(-\infty,\infty)}(x), \qquad (3.12)$$

with $-\infty < \theta < \infty$ and $0 < \tau < \infty$. We say that X has a **normal (Gaussian) distribution with parameters θ and τ^2**, and we write $X \sim n\,(\theta, \tau^2)$.

The **R** function *pnorm( )* can be used (with the *mean* and *sd* arguments corresponding to θ and τ, respectively) to calculate $P(X < b)$ and $P(X < a)$, the difference of which yields $P(a < X < b)$, for $X \sim n\,(\theta, \tau^2)$, as shown in Fig. 3.5.

Note The normal (Gaussian) distribution is the most important continuous probability distribution for continuous random variables and is used extensively in making statistical inference based on sample data. We will return to this feature later in the text.

Mathematical Moment 3 Integration by Parts
Let $F(x)$ and $g(x)$ be functions such that:

(a) $\displaystyle\int_a^b F(x)g(x)dx < \infty$ (a could be $-\infty$ and b could be ∞)

(b) $g(x)$ has an antiderivative $G(x)$ for all $x \in (a, b)$

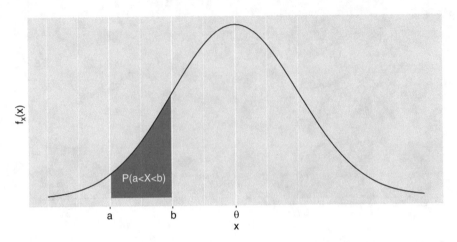

Fig 3.5   Calculating $P(a < X < b)$ for $X \sim n\,(\theta, \tau^2)$

(c) $F(x)$ has a derivative $f(x)$ for all $x \in (a,b)$

and

(d) $\displaystyle\int_a^b G(x)f(x)dx \; < \; \infty$

Then

$$\int_a^b F(x)g(x)dx = F(x)G(x)\Big|_{x=a}^{x=b} - \int_a^b G(x)f(x)dx. \tag{3.13}$$

Example 3.21 Integration by Parts
We can use integration by parts to evaluate integrals with no obvious antiderivatives for their integrands. For example, suppose we wish to evaluate $\int_1^2 x \ln(x)dx$. Taking $F(x) = \ln(x)$ and $g(x) = x$, we have $f(x) = 1/x$ and $G(x) = x^2/2$. Conditions (a)–(d) are all valid, so that it follows from integration by parts that

$$\int_1^2 x\ln(x)dx = \tfrac{x^2}{2}\ln(x)\Big|_{x=1}^{x=2} - \int_1^2 \tfrac{1}{x}\left(\tfrac{x^2}{2}\right)dx = \tfrac{2^2}{2}\ln(2) - \tfrac{1^2}{2}\ln(1) - \int_1^2 \tfrac{x}{2}dx$$

$$= 2\ln(2) - \tfrac{x^2}{4}\Big|_{x=1}^{x=2} = 2\ln(2) - \tfrac{(4-1)}{4} = 2\ln(2) - \tfrac{3}{4}.$$

Mathematical Moment 4 L'Hospital's Rule

Suppose that either

$$\text{(a) Case 1}: \lim_{x \to c} f(x) = \lim_{x \to c} g(x) = 0$$

holds or

$$\text{(b) Case 2}: \lim_{x \to c} \mid g(x) \mid = \infty$$

holds. Then, if both $f'(x)$ and $g'(x)$ exist over the range of interest, it follows that

$$\lim_{x \to c} \frac{f(x)}{g(x)} = \lim_{x \to c} \frac{f'(x)}{g'(x)}, \tag{3.14}$$

provided the latter limit exists.

Example 3.22 Case 1 of L'Hospital's Rule

Suppose we wish to evaluate $\lim_{x \to 0} \frac{e^x - 1}{x}$. Then Case 1 pertains with $f(x) = e^x - 1$ and $g(x) = x$. Since $f'(x) = e^x$ and $g'(x) = 1$, the appropriate derivatives exist, and it follows from L'Hospital's Rule that

$$\lim_{x \to 0} \frac{e^x - 1}{x} = \lim_{x \to 0} \frac{e^x}{1} = 1.$$

Note As is often the case in mathematics, such results can often be obtained in a variety of ways. For example, we can also use the Taylor series expansion of e^x about $x = 0$ (i.e., the Maclaurin series) to evaluate the limit in Example 3.22, as follows:

The Maclaurin series for $f(x) = e^x$ is $1 + x + \frac{x^2}{2!} + \frac{x^3}{3!} + \ldots$ Using this series expansion, we have

$$\frac{e^x - 1}{x} = \frac{\sum_{t=0}^{\infty} \frac{x^t}{t!} - 1}{x} = \frac{\sum_{t=1}^{\infty} \frac{x^t}{t!}}{x} = \sum_{t=1}^{\infty} \frac{x^{t-1}}{t!}.$$

It follows that

$$\lim_{x \to 0} \frac{e^x - 1}{x} = \lim_{x \to 0} \sum_{t=1}^{\infty} \frac{x^{t-1}}{t!} = 1 + \lim_{x \to 0} \sum_{t=2}^{\infty} \frac{x^{t-1}}{t!} = 1 + 0 = 1.$$

Example 3.23 Case 2 of L'Hospital's Rule

Suppose we wish to evaluate $\lim_{x\to\infty} x^2 e^{-x} = \lim_{x\to\infty} \frac{x^2}{e^x}$. Then Case 2 pertains with $f(x) = x^2$ and $g(x) = e^x$. Since $f'(x) = 2x$ and $g'(x) = e^x$, the appropriate derivatives exist, and it follows from L'Hospital's Rule that

$$\lim_{x\to\infty} x^2 e^{-x} = \lim_{x\to\infty} \frac{x^2}{e^x} = \lim_{x\to\infty} \frac{2x}{e^x}(\text{Case 2}) = \lim_{x\to\infty} \frac{2}{e^x}(\text{Case 2}) = 0.$$

(Note that this also shows that $\lim_{x\to\infty} xe^{-x} = 0$.)

Example 3.24 Integration by Parts and L'Hospital's Rule

Integration by parts and L'Hospital's rule are often used together to establish needed results. Consider the integral $\int_0^\infty x^2 e^{-x} dx$ (which will become relevant in Chap. 4). Using integration by parts with $F(x) = x^2$ and $g(x) = e^{-x}$, we see that

$$\int_0^\infty x^2 e^{-x} dx = -x^2 e^{-x}\Big|_0^\infty + \int_0^\infty 2x e^{-x} dx.$$

From Example 3.23, we know that $-x^2 e^{-x}\big|_0^\infty = 0$, so that

$$\int_0^\infty x^2 e^{-x} dx = \int_0^\infty 2x e^{-x} dx.$$

Using integration by parts once more with $F(x) = x$ and $g(x) = e^{-x}$, we obtain (again using Example 3.23 as well)

$$\int_0^\infty 2x e^{-x} dx = -2x e^{-x}\Big|_0^\infty + 2\int_0^\infty e^{-x} dx = 0 - 2e^{-x}\Big|_0^\infty = 2.$$

Combining these results, we have

$$\int_0^\infty x^2 e^{-x} dx = 2.$$

Once we have obtained a bit more familiarity with probability distributions, we will see later in the text that we can obtain this result (and similar ones) without actually evaluating an integral at all!

Mathematical Moment 5 Mathematical Induction

Let A(n) denote a proposition associated with the positive integer n. Suppose we can show that:

(a) A(1) is true

and

(b) If A(*n*) is true, then A(*n*+1) is also true, for arbitrary *n*.

Then it follows that A(*n*) is true for every positive integer *n*. This is called the **Principle of Mathematical Induction**.

Example 3.25 Sum of Squares of First *n* Integers
We have been asked to prove that

$$\sum_{i=1}^{n} i^2 = \frac{n(n+1)(2n+1)}{6}. \tag{3.15}$$

Proof Let A(*n*) denote the equality in (3.15) for integer *n*. Then, A(1) is true, since $\sum_{i=1}^{1} i^2 = 1 = \frac{1(1+1)(2(1)+1)}{6}$. Now, assume that A(*n*) is true for an arbitrary *n*. Then we have:

$$\sum_{i=1}^{n+1} i^2 = \sum_{i=1}^{n} i^2 + (n+1)^2 = \frac{n(n+1)(2n+1)}{6} + (n+1)^2 \text{(Inductive Step)}$$

$$= \frac{n(n+1)(2n+1) + 6(n+1)^2}{6} = \frac{n+1}{6} \left[2n^2 + n + 6n + 6 \right]$$

$$= \frac{n+1}{6} \left[(n+2)(2n+3) \right] = \frac{(n+1)((n+1)+1)(2(n+1)+1)}{6},$$

which establishes that A(*n*+1) is also true. Thus, by the Principle of Mathematical Induction, the equality in (3.15) holds for all *n*. ∎

Example 3.26 Binomial Expansion
Let *a* and *b* be arbitrary constants. Show that

$$(a+b)^n = \sum_{i=0}^{n} \binom{n}{i} a^i b^{n-i}, \text{ for every positive integer } n. \tag{3.16}$$

Proof Let A(*n*) denote the equality in (3.16) for an arbitrary integer *n*. Then,

$$\sum_{i=0}^{1} \binom{1}{i} a^i b^{1-i} = \binom{1}{0} a^0 b^{1-0} + \binom{1}{1} a^1 b^{1-1} = b + a = (a+b)^1,$$

so that A(1) is true. Now assume A(*n*) is true for an arbitrary *n*, so that

$$(a+b)^{n+1} = (a+b)(a+b)^n = (a+b)\sum_{i=0}^{n} \binom{n}{i} a^i b^{n-i} \text{(Inductive Step)}$$

$$= \sum_{i=0}^{n} \binom{n}{i} a^{i+1} b^{n-i} + \sum_{i=0}^{n} \binom{n}{i} a^i b^{n+1-i}$$

$$= \left[\binom{n}{0} a^1 b^n + \binom{n}{1} a^2 b^{n-1} + \ldots + \binom{n}{n-1} a^n b^1 + \binom{n}{n} a^{n+1} b^0 \right]$$

$$+ \left[\binom{n}{0} a^0 b^{n+1} + \binom{n}{1} a^1 b^n + \ldots + \binom{n}{n-1} a^{n-1} b^2 + \binom{n}{n} a^n b^1 \right]$$

$$= \binom{n}{0} b^{n+1} + \left[\binom{n}{1} + \binom{n}{0} \right] ab^n + \left[\binom{n}{1} + \binom{n}{2} \right] a^2 b^{n-1} + \ldots$$

$$+ \left[\binom{n}{n-1} + \binom{n}{n-2} \right] a^{n-1} b^2 + \left[\binom{n}{n} + \binom{n}{n-1} \right] a^n b^1 + \binom{n}{n} a^{n+1}$$

$$= b^{n+1} + (n+1)ab^n + \frac{n(n+1)}{2} a^2 b^{n-1} + \ldots$$

$$+ \frac{n(n+1)}{2} a^{n-1} b^2 + (n+1) a^n b^1 + a^{n+1}$$

$$= \sum_{i=0}^{n+1} \binom{n+1}{i} a^i b^{(n+1)-i},$$

which establishes that A(n+1) is also true. Thus, by the Principle of Mathematical Induction the equality in (3.16) holds for all n. ∎

Mathematical Moment 6 Bonferroni's Inequality

Let $A_1, \ldots, A_n$ be arbitrary events. Then

$$P\left(\bigcap_{i=1}^{n} A_i \right) \geq 1 - \sum_{i=1}^{n} P(A_i^c). \tag{3.17}$$

Proof The proof proceeds via induction. Let A(n) denote the inequality in (3.17) for an arbitrary integer n. A(1) is clearly true, since

$$P\left(\bigcap_{i=1}^{1} A_i \right) = P(A_i) = 1 - P(A_1^c).$$

Now assume that A(n) is true for an arbitrary n, so that

$$P\left(\bigcap_{i=1}^{n} A_i \right) \geq 1 - \sum_{i=1}^{n} P(A_i^c).$$

Then, it follows that

$$P\left(\bigcap_{i=1}^{n+1} A_i\right) = P\left(\bigcap_{i=1}^{n-1} A_i \cap (A_n \cap A_{n+1})\right)$$

$$\geq 1 - \left[\sum_{i=1}^{n-1} P(A_i^c) + P[(A_n \cap A_{n+1})^c]\right] \text{(Inductive Step)}$$

$$= 1 - \sum_{i=1}^{n-1} P(A_i^c) - P(A_n^c \cup A_{n+1}^c).$$

But we know from the property about probabilities of unions that

$$-P\left(A_n^c \cup A_{n+1}^c\right) = -P\left(A_n^c\right) - P\left(A_{n+1}^c\right) + P\left(A_n^c \cap A_{n+1}^c\right) \geq -P\left(A_n^c\right) - P\left(A_{n+1}^c\right),$$

since $P\left(A_n^c \cap A_{n+1}^c\right) \geq 0$. This implies that

$$P\left(\bigcap_{i=1}^{n+1} A_i\right) \geq 1 - \sum_{i=1}^{n-1} P(A_i^c) - P(A_n^c) - P(A_{n+1}^c)$$

$$= 1 - \sum_{i=1}^{n+1} P(A_i^c),$$

so that A($n+1$) is also true. Thus, by the Principle of Mathematical Induction, Bonferroni's Inequality holds for all n. ■

Example 3.27 Using Pairwise Probability Information About Two Variables to Make a Probability Statement About Four Variables
Let X, Y, W, and Z be four random variables such that

$$P(X > Y) = .7, \quad P(Y > W) = .8, \text{ and } P(W > Z) = .9.$$

Then, using Bonferroni's Inequality, we have that

$$\begin{aligned}
P(X > Y > W > Z) &= P(\{X > Y\} \cap \{Y > W\} \cap \{W > Z\}) \\
&\geq 1 - [P(\{X > Y\}^c) + P(\{Y > W\}^c) + P(\{W > Z\}^c)] \\
&= 1 - [(1 - .7) + (1 - .8) + (1 - .9)] = 1 - .6 = .4.
\end{aligned}$$

3.4   Exercises

3.1. Find the Maclaurin series expansion for the function $f(x) = (1 - x)^{-1}$ for $|x| < 1$. Does this series look familiar? Why must we restrict the domain for this series expansion to $|x| < 1$?

3.2. Show that the Maclaurin series expansion for any polynomial is just the polynomial itself.

3.3. Let $X \sim$ Gamma (α, β) with p.d.f. $f_X(x)$. Show that

$$\int_0^\infty f_X(x)dx = 1.$$

3.4. Use an appropriate Taylor series expansion to show that

$$\lim_{x\to\infty} xe^{-x} = \lim_{x\to\infty} x^2 e^{-x} = 0.$$

State and prove a more general result like this for an arbitrary polynomial of degree n; that is, show that

$$\lim_{x\to\infty}\left(\sum_{i=0}^n a_i x^i\right) e^{-x} = 0,$$

where $a_0, \ldots, a_n$ are arbitrary constants.

3.5. Show that $\lim\limits_{x\to\infty} \frac{e^x + e^{-x} - 2}{3x^2} = \frac{1}{3}$.

3.6. Show that $\lim\limits_{x\to\infty} \frac{\ln(1+e^{2x})}{x} = 2$.

3.7. Show that $\lim\limits_{x\to\infty} \ln\left(x^{\frac{1}{x}}\right) = 0$. What does this imply about $\lim\limits_{x\to\infty} x^{\frac{1}{x}}$?

3.8. Show that $\sum\limits_{i=1}^n i = \frac{n(n+1)}{2}$.

3.9. Find $\lim\limits_{x\to 0} \frac{\sin(x) - x}{x^3}$.

3.10. Find $\lim\limits_{x\to 0} \frac{[e^x - 1]^2}{5x^2}$.

3.11. Obtain the Taylor series expansion of the function $f(x) = e^x + e^{-x}$ about $x = 0$.

3.12. Evaluate $\lim\limits_{x\to 0}\left[\frac{x+1-\sin x-\cos x}{x^2+x^3}\right]$.

3.13. Obtain the Taylor series expansion of the function $f(x) = \sin x \cos x$ about $x = 0$.

3.14. Let X be the number of successes obtained in $n_1 > 5$ independent Bernoulli trials each with probability of success p_1.

(a) What is the probability function for X?

Let Y be the number of successes obtained in $n_2 > 5$ independent Bernoulli trials each with probability of success p_2. Assume that all n_2 of this second set of Bernoulli trials are also independent of all n_1 of the first set of Bernoulli trials.

(b) What is the probability that we have exactly three successes in the first set of n_1 trials and exactly five successes in the second set of n_2 trials?

(c) Write down a summation expression for the probability that we get the same number of successes in the first n_1 trials as we do in the second n_2 trials.

(d) Now suppose that $n_1 = n_2 = n$. Find a closed form expression (not just a summation) for the probability that the two sets of n trials have EXACTLY the same number of successes.

3.15. Consider the following experiment. You and your friend each toss a fair coin simultaneously until you both have the same outcome on the same flip (either both heads or both tails). Let X be the number of flips necessary for this to occur. What is the probability function for X?

3.16. The following "facts" are known: 80% of new teenage drivers pass their driver's license test (written and driving) on their first attempt, 15% of new teenage drivers pass their driver's license test on the second attempt, and the remaining 5% of new teenage drivers require at least three attempts before they pass their driver's license test. If a new teenage driver passes her driver's license test on the first attempt, she has probability .001 of having an accident every time she drives a car. If it takes two times for a new teenage driver to pass her driver's license test, she has probability .01 of having an accident every time she drives a car. If it takes more than two times for a new teenage driver to pass her driver's license test, she has probability .05 of having an accident every time she drives a car. Assume throughout this problem that the different times a person drives a car are independent events.

(a) What is the probability that a randomly selected new teenage driver passed her driver's license test on her first attempt?

(b) What is the probability that a randomly selected new teenage driver has an accident the first time she drives after obtaining her license?

(c) If the randomly selected new teenage driver passed her driver's license test on the second attempt, how many accidents should we expect her to have in the first 100 times she drives after obtaining her license?

(d) If the randomly selected new teenage driver passed her driver's license test on the first try, what is the probability that she does not have her first accident until after she has driven more than 50 times?

(e) Suppose the randomly selected new teenage driver has two accidents in her first four driving trips after obtaining her license. What is the probability that it took more than two attempts for her to pass the driver's license test?

3.17. A six-sided die is loaded in such a way that the probability of any particular face resulting from a roll of the die is directly proportional to the number on that face. Let X denote the number observed for a single roll of this die.

(a) What is the probability function for X?

(b) What is the probability that an even number is observed on a single roll of this die?

(c) Suppose we roll this die five times. What is the probability that the same number occurs on each of the five rolls?

3.18. A bag contains five blue balls and three red balls. Someone draws a ball, and then a second ball (with replacement between draws), and, finally, a third ball (again with replacement between draws). Let X denote the number of red balls obtained in the three balls drawn from the bag. What is the probability function for X?

3.19. Consider the experiment described in Exercise 3.18, except now do not replace the balls between draws. What is the probability function for X under these conditions?

3.20. Consider once again the experiment described in Exercise 3.18, except now replace the first ball drawn before drawing the second ball, but do not replace the second ball drawn before drawing the third ball. What is the probability function for X under these conditions?

3.21. Consider one last time the experiment described in Exercise 3.18, except now do not replace the first ball drawn before drawing the second ball but do replace the second ball drawn before drawing the third ball. What is the probability function for X under these conditions? Compare the probability functions obtained in Exercises 3.18–3.21.

3.22. Show that $\sum_{i=1}^{n} i = n(n+1)/2$ without using induction.

3.23. Find the form of the Taylor series expansion of $f(x) = \ln(1+x)$ about $x = 0$ for $-1 < x < \infty$.

3.24. Use a Taylor series expansion to show that an approximate solution to the equation $\sin x = \lambda x$ near $x = \pi$ is given by $x = \pi/(1 - \lambda)$.

3.25. Let $\Gamma(\cdot)$ denote the gamma function.

(a) For any $t > 1$, show that $\Gamma(t) = (t-1)\Gamma(t-1)$.
(b) Show that $\Gamma(n) = (n-1)!$ for any positive integer n.

3.26. Let $\mu > 0$ be arbitrary. Show that

$$\int_{\mu}^{\infty} [\Gamma(k)]^{-1} z^{k-1} e^{-z} \, dz = \sum_{x=0}^{k-1} \frac{\mu^x e^{-\mu}}{x!}$$

for every positive integer k. Discuss how this equation connects the gamma distribution and the Poisson distribution.

3.27. Let λ and d be constants not depending on n. Show that

$$\lim_{n \to \infty} \left[1 + \frac{\lambda}{n} + o\left(\frac{1}{n}\right) \right]^{dn} = e^{\lambda d},$$

where $o\left(\frac{1}{n}\right)$ as $n \to \infty$ means that $\lim_{n \to \infty} n \, o\left(\frac{1}{n}\right) = 0$.

3.28. Show that $\int_0^\pi e^x \sin(x)\, dx = (e^\pi + 1)/2$. (Do *not* use integral tables to find the antiderivative of $e^x \sin(x)$.)

3.29. Show that $\sum_{i=1}^n i^3 = n^2(n+1)^2/4$.

3.30. Six towns lie on a specific stretch of a river. Suppose that the dam structure on this river is such that three of these towns each have probability .05 that they will be flooded at least once in a given year, while the other three towns are more secure, and each has probability only .025 that they will be flooded at least once in a given year. Show that the probability is at least .775 that none of these towns will be flooded in a given year.

3.31. The probability is 0.26 that a child of two left-handed parents will also be left-handed. If a couple who are both left-handed have four children, find the probability distribution for the number of their children who are also left-handed. Assume independence of this trait between births of the children.

3.32. Consider the random variable that counts the number of heads in 12 flips of a fair coin. Find the probability of getting

(a) Exactly 3 heads
(b) At least 7 heads
(c) Fewer than 11 heads
(d) More than 6 tails
(e) At most 4 heads

3.33. Suppose $X \sim Binom(25, .3)$. Use the **R** function *dbinom( )* to find the complete probability distribution for X. Graph this probability distribution and comment on the shape and at least one other feature of the distribution.

3.34. *USA Today* (2013) reported on a number of results from a survey conducted by the Public Religion Research Institute in the United States in 2013. One of the questions asked in the survey was: "True or False: God plays a direct role in determining which team wins a sporting event." Twenty seven percent of the respondents answered "True", indicating that they believe that God does play a direct role in determining winners of sporting events. Suppose you randomly select 15 individuals and ask them this same question. Let X denote the number of individuals in your poll who indicate that they believe that God plays a direct role in determining winners of sporting events. If the results of the Public Religion Research Institute survey are applicable to your experiment, find:

(a) A general expression for the probability distribution for X
(b) $P(X < 4)$
(c) $P($no respondents in your poll indicated that they believe that God plays a direct role in determining winners of sporting events$)$
(d) $P($at least half of the respondents in your poll indicated that they believe that God plays a direct role in determining winners of sporting events$)$

3.35. In a 2007 survey, the Pew Research Center (January, 2010) found that among marriages between individuals with opposite genders, 53% of spouses had the same educational level, 19% of men had more education than their spouses, and 28% of women had more education than their spouses. Suppose you are in a class with 15 men and 20 women (including yourself) and assume that all 35 of you will marry (at least once!) someone of the opposite gender.

(a) What is the probability that more than half of the men in your class will marry (for the first time) a woman with more education?
(b) What is the probability that more than half of the women in your class will marry (for the first time) a man with more education?
(c) What is the probability that more than half of your class (men and women combined) will marry (for the first time) someone of the opposite gender with the same educational level?

3.36. In the game of Monopoly, a player must stay in jail until they roll doubles with a pair of fair dice (assume they do not have a Get Out of Jail Free card and cannot pay a fine to get out). What is the probability:

(a) The player will get out of jail on her first roll?
(b) The player will not get out of jail until after her fifth roll?
(c) The player will get out of jail before her third roll?

3.37. Meilman et al. (1998) reported on the regional distribution of students who carry weapons on college campuses, using data obtained from the nationwide Core Alcohol and Drug Survey of 24,545 students at 61 institutions during the 1994–1995 academic years. Dividing the United States into four regions, they found the following information for the number of US college students who reported carrying a weapon at least once in the previous 30 days:

Region	Carried weapon	Did not carry weapon	Total
Northeast	262	4335	4597
North Central	559	7547	8106
South	451	4197	4648
West	446	6748	7194

Assume that these sample percentages apply exactly to the students on your campus (pick the region) and that you are enrolled in a class with 100 other students. If you decide to interview ten students (not yourself!) randomly selected (without replacement between interviews) from your class, find the following:

(a) Probability that none of the students in your interviewed group has carried a weapon at least once in the past 30 days
(b) Probability that more than half of the students in your interviewed group have carried a weapon at least once in the past 30 days
(c) Probability that at least one individual in your interviewed group has carried a weapon at least once in the past 30 days

(d) Probability that less than three of the students in your interviewed group have carried a weapon at least once in the past 30 days

(e) Probability that exactly half of the individuals in your interviewed group have carried a weapon at least once in the past 30 days

Do you think the percentages have changed since this survey in 1994–1995?

3.38. A student intends to stand in the middle of campus and randomly ask individuals walking by to complete a survey. She plans to continue to ask individuals to complete the survey until ten surveys have been completed, at which time she will stop. You can assume that individuals agree (or decline) to complete the survey independently of each other. Assume that the probability any individual will complete the survey is 0.20.

(a) What is the probability that the 30th individual the student asks to complete the survey is the last person that the student will have to ask?

(b) What is the probability that the first ten individuals the student asks all complete the survey?

(c) What is the probability that the student will have to ask more than 20 individuals before she is able to stop?

(d) Suppose that seven of the first ten individuals the student asks agree to complete the survey. What is the probability that she will not need to ask any more than five additional individuals to complete the survey?

3.39. Suppose $X \sim Binom$ (20, .6).

(a) Use the **R** function *dbinom( )* to obtain the full probability distribution for X and comment on the important features of the distribution.

(b) Use the calculations from part (a) to find $P(7 \leq X \leq 15)$. Find this same probability by using the cumulative **R** function *pbinom( )*.

(c) Use the cumulative R function *pbinom( )* to obtain $P(X < 7)$.

(d) What do the answers to parts (b) and (c) say about $P(X > 15)$? Find this same probability directly using only the cumulative **R** function *pbinom( )*.

3.40. Suppose $X \sim NegBin$ (8, .4).

(a) Use the **R** function *dnbinom( )* to find $P(X = x)$ for $x \in \{8, 9, \ldots, 15, 16\}$.

(b) Find $P(10 < X \leq 14)$ using the individual calculations from part (a). Find this same probability using the cumulative **R** function *pnbinom( )*.

(c) What does this say about $P(X < 8)$ and $P(X > 16)$?

3.41. Suppose $X \sim Poisson$ (5).

(a) Use the **R** function *dpois( )* to find $P(X = x)$ for $x \in \{4, 5, \ldots, 16, 17\}$.

(b) Find $P(5 \leq X \leq 15)$ using the individual calculations from part (a). Find this same probability using the cumulative **R** function *ppois( )*.

(c) Use the cumulative **R** function *ppois( )* to obtain $P(X > 17)$?

(d) What do the answers to parts (a) and (c) say about $P(0 \leq X < 4)$? Find this same probability directly using only the cumulative **R** function *ppois( )*.

3.42. Suppose X has a hypergeometric distribution with parameters $N = 50$, $b = 17$, and $n = 20$.

(a) What is the support for X?
(b) Use the **R** function *dhyper*( ) to obtain the full probability distribution for X and comment on the important features of the distribution.
(c) Find $P(5 < X \leq 12)$ using the individual calculations from part (a). Find this same probability using the cumulative **R** function *phyper*( ).
(d) Use the cumulative **R** function *phyper*( ) to obtain $P(X \geq 13)$.
(e) What do the answers to parts (c) and (d) say about $P(X \leq 5)$?

3.43. Suppose X has a hypergeometric distribution with parameters $N = 35$, $b = 7$, and $n = 10$.

(a) What is the support for X?
(b) Use the **R** function *dhyper*( ) to obtain the full probability distribution for X and comment on the important features of the distribution.
(c) Find $P(5 < X \leq 10)$ using the individual calculations from part (a). Find this same probability using the cumulative **R** function *phyper*( ).
(d) Use the cumulative **R** function *phyper*( ) to obtain $P(X \leq 3)$.
(e) Use the cumulative **R** function *phyper*( ) to find $P(2 \leq X < 6)$.

3.44. Suppose X has a hypergeometric distribution with parameters $N = 40$, $b = 25$, and $n = 20$.

(a) What is the support for X?
(b) Use the **R** function *dhyper*( ) to obtain the full probability distribution for X and comment on the important features of the distribution.
(c) Find $P(12 \leq X < 17)$ using the individual calculations from part (a). Find this same probability using the cumulative **R** function *phyper*( ).
(d) Use the cumulative **R** function *phyper*( ) to obtain $P(X \geq 13)$.
(e) Use the cumulative **R** function *phyper*( ) to find $P(X > 11)$.

3.45. Let $X \sim Geom\ (.3)$.

(a) Use the **R** function *dgeom*( ) to find $P(X = x)$ for $x \in \{7, 8, \ldots, 15, 16\}$.
(b) Find $P(10 < X \leq 15)$ using the individual calculations from part (a). Find this same probability using the cumulative **R** function *pgeom*( ).
(c) Use the cumulative **R** function *pgeom*( ) to obtain $P(X < 5)$.
(d) Use the cumulative **R** function *pgeom*( ) to calculate $P(X \geq 11)$.

3.46. Let $X \sim Gamma\ (5, 3)$. Use an appropriate **R** function to find the following probabilities:

(a) $P(2 < X < 5)$
(b) $P(X > 15)$
(c) $P(X < 9)$
(d) $P(X \geq 15)$

3.47. Let $X \sim Exp(7)$. Use an appropriate **R** function to find the following probabilities:

(a) $P(X < 12)$
(b) $P(9 \leq X < 15)$
(c) $P(X \geq 11)$
(d) $P(X \leq 11)$

3.48. Let $X \sim \chi^2(11)$. Use an appropriate **R** function to find the following probabilities:

(a) $P(7 \leq X \leq 13)$
(b) $P(X < 8)$
(c) $P(X \geq 12)$

3.49. Let $X \sim \chi^2(8)$ and let $Y \sim \chi^2(14)$. Use an appropriate **R** function to compare the following probabilities:

(a) $P(X > 10)$ and $P(Y > 10)$
(b) $P(X < 12)$ and $P(Y < 12)$
(c) $P(5 < X < 9)$ and $P(5 < Y < 9)$
(d) What do these calculations tell you about the effect that the degrees of freedom has on the chi-square distribution?

3.50. Let $X \sim n(5, 12)$. Use an appropriate **R** function to find the following probabilities:

(a) $P(X > 6.2)$
(b) $P(5.1 < X < 9.7)$
(c) $P(X \leq 7.4)$

3.51. Let $X \sim n(6, 9)$ and let $Y \sim n(12, 9)$. Use an appropriate **R** function to compare the following probabilities:

(a) $P(3 < X < 9)$ and $P(3 < Y < 9)$.
(b) $P(X > 9)$ and $P(Y > 9)$.
(c) $P(X > 6)$ and $P(Y > 12)$.
(d) $P(X < 9)$ and $P(Y < 15)$
(e) Discuss these results in the context of what effect the two parameters θ and τ^2 have on the normal distribution.

3.52. Let $X \sim n(6, 9)$ and let $Y \sim n(6, 16)$. Use an appropriate **R** function to compare the following probabilities:

(a) $P(3 < X < 9)$ and $P(3 < Y < 9)$.
(b) $P(X > 9)$ and $P(Y > 9)$.
(c) $P(X > 6)$ and $P(Y > 12)$.
(d) $P(X < 9)$ and $P(Y < 15)$.
(e) Discuss these results in the context of what effect the two parameters θ and τ^2 have on the normal distribution.

3.53. National Public Radio (2014) reported on a number of results from a survey conducted by the National Science Foundation in the United States in 2012. One of the questions asked in the survey was "Does the earth revolve around the sun, or does the sun revolve around the earth?" Twenty six percent of the respondents said that the sun revolved around the earth! Suppose you randomly select 20 individuals and ask them this question. If the results of the National Science Foundation survey are applicable:

(a) What is the probability that at least one of the individuals you interview believes that the sun revolves around the earth?
(b) What is the probability that you will not find someone who believes that the sun revolves around the earth before your seventh interview?
(c) What is the probability that exactly half of the individuals you interview believe that the sun revolves around the earth?
(d) What is the probability that less than half of the individuals you interview believe that the sun revolves around the earth?
(e) What is the probability that more than half of the individuals you interview believe that the sun revolves around the earth?

3.54. The Business Insider (2012) reported on a number of results from a national survey conducted by Wakefield Research (commissioned by Citrix) in August 2012. One of the questions asked in the survey was "Can stormy weather interfere with cloud computing?" Fifty one percent of the respondents (including a majority of millennials) agreed that stormy weather can interfere with cloud computing! Suppose you randomly select 30 individuals and ask them this same question. If the results of the Wakefield Research survey are applicable:

(a) What is the probability that no more than ten of the individuals you interview believe that stormy weather can interfere with cloud computing?
(b) What is the probability that between 20% and 40%, inclusive, of the individuals you interview believe that stormy weather can interfere with cloud computing?
(c) What is the probability that less than 50% of the individuals you interview believe that stormy weather can interfere with cloud computing?

3.55. How does attending church services on a regular basis (at least once a week) affect your approval of using torture against suspected terrorists? The Cable News Network (2009) reported on a number of results obtained in a survey conducted April 14–21, 2009, by the Pew Research Center. One of the questions asked in the survey was "Do you think the use of torture against suspected terrorists in order to gain important information can often be justified, sometimes be justified, rarely be justified, or never be justified?" Forty two percent of the respondents who "seldom or never" attend church services agreed that the use of torture against suspected terrorists is "often" or "sometimes" justified. On the other hand, 54% of the respondents who attend church services at least once a week agreed that the use of torture against suspected terrorists is "often" or "sometimes" justified. Suppose that these results of the Pew Research Center are applicable.

(a) If you select a random sample of 50 individuals who "seldom or never" attend church services, what is the probability that less than half of the individuals you interview will agree that the use of torture against suspected terrorists is "often" or "sometimes" justified?
(b) If you select a random sample of 50 individuals who attend church services on a regular basis (at least once a week), what is the probability that more than half of the individuals you interview will agree that the use of torture against suspected terrorists is "often" or "sometimes" justified?

3.56. Several years ago, an elementary school in New York state reported that its incoming kindergarten class contained five sets of twins. This was given statewide press, with a quote from the principal that this was a "statistical impossibility". Was it really? The probability of a twin birth is roughly 1/90. Assume that a typical entering kindergarten class at a school has approximately 60 children and count each pair of twins as only a single birth so that the 60 children represent 60 distinct births.

(a) What is the probability distribution for the number of twin births among 60 distinct births in an incoming kindergarten class?
(b) What is the probability of an event as unusual or more unusual than the incoming class with five sets of twins in the school in New York? Is this "statistically impossible"?
(c) Repeat part (b) if we expand our discussion to include all 62 counties in New York state, assuming each county has five elementary schools with three classes of 20 students each?
(d) Under the expanded setting in part (c), how many of the elementary schools do you think we should have expected to have at least five sets of twins in their incoming kindergarten class?
(e) Repeat part (d) if we expand our sample space further to include all 50 states over each of the past 10 years. For computational purposes, make the (unreasonable) assumption here that each of the 50 states has been similar to the state of New York with regard to both the number and size of kindergarten classes over each of the past 10 years.

3.57. Suppose that the number of chocolate chips in a mass-produced cookie follows a Poisson distribution with mean λ.

(a) Find an expression for the probability, say γ, that a randomly chosen cookie contains at least one chocolate chip.
(b) How large must the mean rate λ be in order to ensure that $\gamma \geq .90$?
(c) Repeat parts (a) and (b) under the more stringent condition of requiring that a randomly chosen cookie contains at least two chocolate chips?

3.58. National Public Radio (2014) reported on a number of results from a survey conducted by the National Science Foundation in the United States in 2012. One of the questions asked in the survey was "True or False: The universe began with a huge explosion." Sixty one percent of the respondents answered False, indicating that they did not believe that the universe began with a huge explosion. Suppose you randomly select individuals and ask them this question. If the results from the National Science Foundation survey are applicable, what is the probability that:

(a) You find nine individuals in the first 15 individuals you interview who do not believe the universe began with a huge explosion?
(b) You do not find a person who believes the universe began with a huge explosion until your seventh interview?
(c) You find the fourth person who does not believe the universe began with a huge explosion on your eighth interview?

3.59. Suppose the time (in seconds, s) taken by a computer on a local area network to establish a connection to a remote site follows a $n\,(15s,\,9s^2)$ distribution.

(a) What proportion of the connections will occur in less than 12.2 s?
(b) What proportion of the connections will occur between 12.2 s and 16.5 s?
(c) What proportion of the connections will take longer than 10 s?
(d) 90% of the connections will occur in less than how many seconds?
(e) 75% of the connections will take longer than how many seconds?

3.60. An automobile manufacturer claims that the fuel consumption for a certain make and model of car should average 28 miles per gallon. After driving such a car for over 6 months (long enough to get through the typical break-in period), you notice that you are only getting 20 miles per gallon. During a phone conversation with a customer service representative for the company, you are told that the standard deviation of the fuel consumption for your car is 3 miles per gallon. Assuming the company's claims are valid, what is the chance that your car would be performing as badly as it is or worse?

3.61. There are 60 faculty members for a Department of English at a major university. As chairperson, you need to select six of these faculty to serve on the Department's Fiscal Committee for the coming year.

(a) How many different Fiscal Committees of size six can you create from your faculty?
(b) If there are 35 female faculty members and 25 male faculty members in the department and you choose the six faculty members for your Fiscal Committee completely at random, what is the probability that your committee will have an equal number of female and male members?
(c) If there are 22 Full Professors, 24 Associate Professors, and 14 Assistant Professors in the Department and you choose the six faculty members for your Fiscal Committee completely at random, what is the probability that the committee will consist of 3 Full Professors, 2 Associate Professors, and 1 Assistant Professor?

3.62. A hand of 13 cards is to be dealt at random and without replacement from an ordinary deck of 52 playing cards (no jokers).

(a) Find the probability that the hand contains at least 2 kings.

(b) Find the conditional probability that the hand contains at least 3 kings given that the hand contains at least 2 kings.
(c) Find the probability that you will have at least nine cards from one suit.
(d) Find the probability that you will have at least six honor cards (ace, king, queen, jack, and ten).

3.63. In a state lottery, you are asked to choose six numbers (without duplication) from a pool of 53 numbers. At the time of the drawing, the state officials also choose six numbers (without duplication) from the pool of 53 numbers. Let X denote the number of matches between your set of six numbers and those selected by the state lottery. Determine the complete probability distribution for X. If the big prizes only go to those who have at least four matches, what are the chances of you being one of them?

3.64. Udias and Rice (1975) showed that a gamma distribution with parameters $\alpha = .509$ and $\beta = 870$ could be used as an approximate model for the number of hours between a sequence of small earthquakes. If we let X denote the number of hours between two such small earthquakes, find the following probabilities:

(a) $P(X < 5 \text{ hours})$
(b) $P(4 \text{ hours} < X < 15 \text{ hours})$
(c) $P(X > 20 \text{ hours})$

3.65. Scores on a standardized (over a very large population) IQ test are approximately distributed as $n\,(100, 225)$. What is the approximate probability that a person achieves a score X satisfying:

(a) $X > 140$?
(b) $X < 80$?
(c) $90 < X < 120$?

3.66. During a laboratory experiment, the number of radioactive particles passing through a detector (counter) can be approximated by the Poisson distribution with an average of five particles passing per millisecond. If we let X denote the number of particles passing through the detector in a given millisecond, find:

(a) $P(X > 9 \text{ particles})$
(b) $P(4 \text{ particles} < X < 12 \text{ particles})$
(c) $P(X > 7 \text{ particles})$

3.67. An electrical firm manufactures light bulbs that have a length of life before burnout (in hours) that can be reasonably approximated by a $n\,(700, 1600)$ distribution. If we let X denote the lifetime (in hours) for a typical bulb from this firm, find the following:

(a) $P(X > 750 \text{ hours})$
(b) $P(X < 650 \text{ hours})$
(c) $P(640 \text{ hours} < X < 760 \text{ hours})$

Chapter 4
General Properties of Random Variables

In Chap. 3, we introduced the concept of a probability distribution for a random variable and discussed a number of specific discrete and continuous probability distributions. In this chapter, we concentrate on important properties of these probability distributions that will help us better understand how they can be used to model experimental studies. Throughout the chapter (and the rest of the text), we will use the notation p.d.f. to refer to both the probability function (discrete variables) and the probability distribution function (continuous variables).

4.1 Cumulative Distribution Function

Definition 4.1 Let X be a random variable with p.d.f. $f_X(x)$. The **cumulative distribution function (c.d.f.)**, $F_X(x)$, **for X** is then

$$F_X(x) = P(X \leq x) = \sum_{t \leq x} f_X(t), \text{ if } X \text{ is discrete,}$$

$$= \int_{-\infty}^{x} f_X(t)dt, \text{ if } X \text{ is continuous.}$$

Note that in the discrete setting, the c.d.f. is a step function with steps at each of the values that have positive probabilities, and the size of a step at a given value is equal to the probability associated with that value. In the continuous setting, the c.d.f. is a continuous function everywhere.

© Springer Nature Switzerland AG 2020
D. Wolfe, G. Schneider, *Primer for Data Analytics and Graduate Study in Statistics*,
https://doi.org/10.1007/978-3-030-47479-9_4

Example 4.1 Discrete Random Variable

Let X be a discrete random variable with p.d.f.

$$\begin{aligned}
f_X(x) &= .1, x = -4, \\
&= .2, x = 0, \\
&= .1, x = 1, \\
&= .4, x = 2, \\
&= .2, x = 5, \\
&= 0, \text{elsewhere.}
\end{aligned}$$

Then the cumulative distribution function (c.d.f.) for X is given by (see Fig. 4.1)

$$\begin{aligned}
F_X(x) &= 0, x < -4, \\
&= .1, -4 \le x < 0, \\
&= .3, 0 \le x < 1, \\
&= .4, 1 \le x < 2, \\
&= .8, 2 \le x < 5, \\
&= 1, x \ge 5.
\end{aligned}$$

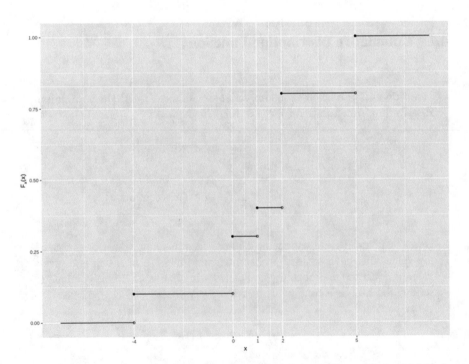

Fig. 4.1 Step function cumulative distribution function (c.d.f.) for the discrete random variable in Example 4.1

Example 4.2 Exponential Distribution

Let X be an exponential random variable with parameter $\beta > 0$ and p.d.f.

$$f_X(x) = \frac{1}{\beta} e^{-\frac{x}{\beta}} I_{(0,\infty)}(x).$$

Then the cumulative distribution function (c.d.f.) for X is given by

$$F_X(x) = \int_0^x f_X(t)dt = \int_0^x \frac{1}{\beta} e^{-\frac{t}{\beta}} dt = -e^{-\frac{t}{\beta}} \Big|_{t=0}^{t=x}$$

$$= 1 - e^{-\frac{x}{\beta}}, 0 \leq x < \infty,$$
$$= 0, \ -\infty < x < 0.$$

See Fig. 4.2 for the case when $\beta = 0.75$.

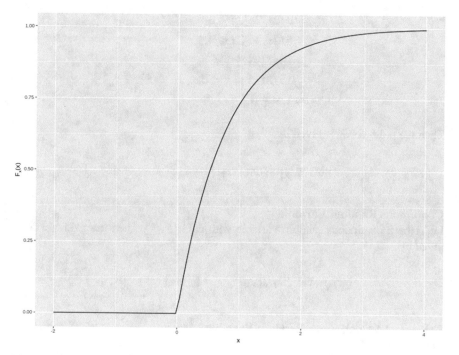

Fig. 4.2 Cumulative distribution function (c.d.f.) for an exponential distribution with parameter $\beta = 0.75$

4.1.1 Relationship Between c.d.f. and p.d.f

Discrete Random Variable

Let X be a discrete random variable with c.d.f. $F_X(x)$ and p.d.f. $f_X(x)$. Then the p.d.f. $f_X(x)$ has positive probabilities associated with each of the jump points (i.e., steps) in the c.d.f. and is given by

$$f_X(t) = P(X = t) = P(X \le t) - P(X < t)$$
$$= F_X(t) - F_X(t^-),$$

where $F_X(t^-) = \lim_{x \uparrow t} F_X(x) =$ left hand limit of $F_X(x)$ at $x = t$.

Example 4.3 Discrete Random Variable

Let X be a discrete random variable with c.d.f.

$$\begin{aligned}
F_X(x) &= 0, x < 1, \\
&= 1/21, 1 \le x < 2, \\
&= 3/21, 2 \le x < 3, \\
&= 6/21, 3 \le x < 4, \\
&= 10/21, 4 \le x < 5, \\
&= 15/21, 5 \le x < 6, \\
&= 1, x \ge 6.
\end{aligned}$$

Then the p.d.f. for X is given by

$$f_X(x) = \frac{x}{21} \, I_{\{1,2,3,4,5,6\}}(x).$$

Continuous Random Variable

Let X be a continuous random variable with c.d.f. $F_X(x)$. Then the p.d.f. for X is given by

$$f_X(x) = \frac{d}{dx} F_X(x).$$

Example 4.4 Continuous Random Variable

Let X have c.d.f.

$$\begin{aligned}
F_X(x) &= 0, \quad x < 0, \\
&= 1 - xe^{-x} - e^{-x}, \quad x \ge 0.
\end{aligned}$$

It follows that

$$f_X(x) = \frac{d}{dx} F_X(x) = -e^{-x} + xe^{-x} + e^{-x} = xe^{-x} I_{(0,\infty)}(x).$$

Thus, X has a *Gamma* $(\alpha = 2, \beta = 1)$ distribution.

Note The c.d.f. uniquely determines the probability distribution for a random variable. Thus, if we can identify a given c.d.f., we can use it to determine the associated probability distribution.

Example 4.5 Gamma Distribution
Suppose $X \sim$ *Gamma* $(1, \beta)$. The c.d.f. for X is

$$F_X(x) = \int_0^x \frac{1}{\beta} e^{-\frac{t}{\beta}} dt = 1 - e^{-\frac{x}{\beta}}, \quad x \geq 0$$

$$= 0, \quad x < 0.$$

Let $Y = X/\beta$. Then the c.d.f. for Y is given by

$$G_Y(y) = P(Y \leq y) = P\left(\frac{X}{\beta} \leq y\right) = P(X \leq \beta y) = F_X(\beta y) = 1 - e^{-\frac{\beta y}{\beta}}$$

$$= 1 - e^{-y}, \quad y \geq 0$$

$$= 0, y < 0.$$

At this point, we could just recognize this to be the c.d.f. for a gamma distribution with parameters $\alpha = \beta = 1$. Of course, since we are still learning how to recognize c.d.f.s, we can always differentiate the c.d.f. $G_Y(y)$ to obtain the p.d.f. $g_Y(y) = \frac{d}{dy} G_Y(y) = e^{-y} I_{[0,\infty)}(y)$, which we do recognize as the p.d.f. for a *Gamma* $(1, 1)$ distribution.

4.1.2   General Properties of a c.d.f. $F_X(x)$

(i) $\lim_{x \to -\infty} F_X(x) = 0$;

(ii) $\lim_{x \to \infty} F_X(x) = 1$;

(iii) For any $a < b$, $F_X(a) \leq F_X(b)$, so that $F_X(x)$ is a nondecreasing function for all x.

(iv) $\lim_{h \to 0^+} F_X(x + h) = F_X(x)$ for all x. Thus, $F_X(x)$ is always continuous from the right, whether X is continuous or discrete.

4.2 Median of a Probability Distribution

Definition 4.2 Let X be a random variable with c.d.f. $F_X(x)$. We say that m is a **median for the X distribution** if

$$F_X(m^-) \leq \frac{1}{2}$$

and

$$F_X(m) \geq \frac{1}{2}.$$

Note If X is continuous, then $F_X(m) = \frac{1}{2}$.

Example 4.6 Continuous Random Variable: Gamma $(1, \beta)$ Distribution
Let X have c.d.f.

$$F_X(x) = 0, \quad x < 0$$
$$= 1 - e^{-x/\beta}, \quad x \geq 0,$$

where $\beta > 0$. Since X is a continuous random variable, we simply set

$$F_X(m) = \frac{1}{2} = 1 - e^{-\frac{m}{\beta}},$$

to obtain

$$e^{-\frac{m}{\beta}} = 1 - \frac{1}{2} = \frac{1}{2} \Rightarrow -\frac{m}{\beta} = -\ln(2) \Rightarrow m = \beta \ln(2).$$

Thus, the median for a *Gamma* $(1, \beta)$ distribution is $m = \beta \ln(2)$.

Example 4.7 Discrete Random Variable
Let X have c.d.f.

$$F_X(x) = 0, x < -4$$
$$= .1, \ -4 \leq x < 0$$
$$= .3, 0 \leq x < 1$$
$$= .4, 1 \leq x < 2$$
$$= .8, 2 \leq x < 5$$
$$= 1, x \geq 5.$$

Since $F_X(2) = .8 \geq .5$ and $F_X(2^-) = .4 \leq .5$, it follows that $m = 2$ is a median for this distribution. [It is also reasonable (and some statisticians do) to claim that all $m \in (1, 2]$ should be considered as multiple medians for this distribution.]

4.3 Symmetric Probability Distribution

Definition 4.3 Let X be a random variable with p.d.f. $f_X(x)$ and c.d.f. $F_X(x)$. We say that the associated probability distribution is **symmetric about the point c** if

$$f_X(c - x) = f_X(c + x) \quad \text{for all } x. \tag{4.1}$$

Continuous Random Variable

If X is continuous and (4.1) holds, then, for any x, we have

$$F_X(c + x) = \int_{-\infty}^{c+x} f_X(t)dt = \int_{-\infty}^{\infty} f_X(t)dt - \int_{c+x}^{\infty} f_X(t)dt$$

$$= 1 - \int_{c+x}^{\infty} f_X(t)dt \underset{y=2c-t}{=} 1 - \int_{-\infty}^{c-x} f_X(2c - y)dy$$

$$= 1 - \int_{-\infty}^{c-x} f_X(y)dy = 1 - F_X(c - x),$$

since, from (4.1), it follows that $f_X(2c - y) = f_X(c + (c - y)) = f_X(c - (c - y)) = f_X(y)$.

Thus, if the distribution of X is continuous and symmetric about the point c, then

$$F_X(c + x) + F_X(c - x) = 1 \quad \text{for all } x. \tag{4.2}$$

This is illustrated in Figs. 4.3 and 4.4.

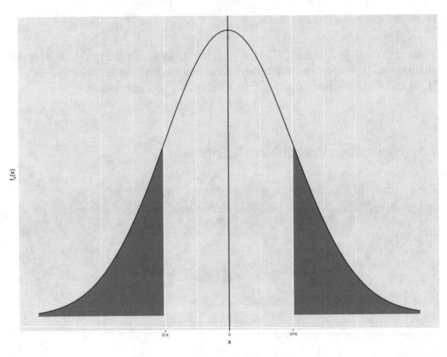

Fig. 4.3 Example of a probability density function (p.d.f.) for a continuous random variable with probability distribution that is symmetric about the point c

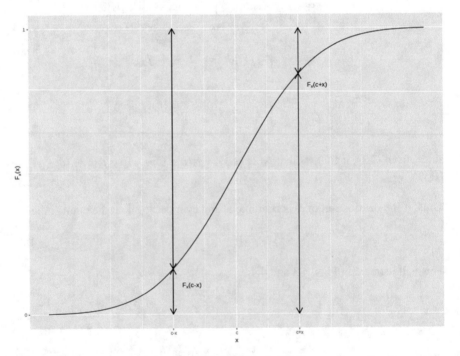

Fig. 4.4 Example of a cumulative distribution function (c.d.f.) for a continuous random variable with probability distribution that is symmetric about the point c

Lemma 4.1 Median for a Symmetric Continuous Distribution

If X has a continuous distribution that is symmetric about c, then c is a median for the distribution.

Proof Taking $x = 0$ in (4.2) yields $F_X(c) = \frac{1}{2}$ and the result follows immediately. ∎

Example 4.8 Normal Distribution

Let $X \sim n\,(\mu, \tau^2)$. Then

$$f_X(x) = \left(2\pi\tau^2\right)^{-\frac{1}{2}} e^{-\frac{1}{2\tau^2}(x-\mu)^2} I_{(-\infty,\infty)}(x)$$

$$\Rightarrow \quad f_X(\mu + t) = \left(2\pi\tau^2\right)^{-\frac{1}{2}} e^{-\frac{1}{2\tau^2}(\mu+t-\mu)^2} = \left(2\pi\tau^2\right)^{-\frac{1}{2}} e^{-\frac{1}{2\tau^2}t^2} = \left(2\pi\tau^2\right)^{-\frac{1}{2}} e^{-\frac{1}{2\tau^2}(\mu-t-\mu)^2} = f_X(\mu - t)$$

$\Rightarrow$ the $n\,(\mu, \tau^2)$ distribution is symmetric about $\mu \Rightarrow \mu$ is the median of the distribution.

Note If X has a discrete distribution that is symmetric about c, then the discrete analogue to (4.2) is

$$F_X(c + x) + F_X((c - x)^-) = 1 \quad \text{for all } x. \tag{4.3}$$

In this discrete setting, it is not necessarily true that $F_X(c) = \frac{1}{2}$, although c is still a median for the X distribution.

Example 4.9 Discrete Distribution Symmetric About c but $F_X(c) \neq \frac{1}{2}$

Consider the probability distribution given by

$$P(X = -1) = P(X = 1) = .4 \quad \text{and} \quad P(X = 0) = .2.$$

This probability distribution is clearly symmetric about $c = 0$, but $F_X(0) = .6$, not 1/2. However, 0 is still a median, since $F_X(0) = .6 \geq .5$ and $F_X(0^-) = .4 \leq .5$.

4.4 Mathematical Expectations

Definition 4.4 Let X be a random variable with p.d.f. $f_X(x)$ and let $g(X)$ be some function of X. The **expected value of** $g(X)$ is defined by

$$E[g(X)] = \int_{-\infty}^{\infty} g(x) f_X(x)\, dx \quad \text{if } X \text{ is continuous}$$

or

$$E[g(X)] = \sum_x g(x) f_X(x) \quad \text{if } X \text{ is discrete}$$

Example 4.10 Continuous Variable

Let X be a continuous random variable with p.d.f.

$$f_X(x) = \frac{1}{2\sqrt{x}} I_{(0,1)}(x).$$

Then

$$E[X] = \mu_X = \int_0^1 \frac{x}{2\sqrt{x}} dx = \frac{x^{\frac{3}{2}}}{2\left(\frac{3}{2}\right)} \Big|_{x=0}^{x=1} = \frac{1}{3}$$

and

$$E\left[\sqrt{X}\right] = \int_0^1 \frac{\sqrt{x}}{2\sqrt{x}} dx = \int_0^1 \frac{1}{2} dx = \frac{x}{2} \Big|_{x=0}^{x=1} = \frac{1}{2}.$$

Example 4.11 Geometric Variable

Flip a fair coin until the first head occurs. Let X be the total number of flips required. Then X has a geometric distribution with $p = 0.5$, and its p.d.f. is given by

$$p_X(x) = \left(\frac{1}{2}\right)^x I_{\{1,2,3,\dots\}}(x).$$

Consider the following game:

$$\text{Win } \$2^X \text{ if } X \leq 19$$

$$\text{Win } \$2^{20} \text{ if } X \geq 20.$$

Suppose you are asked to pay $20 to play the game. Should you? Our expected winnings are given by $E[u(X)]$, where

$$u(x) = 2^x, \quad \text{if } x \leq 19$$
$$= 2^{20}, \quad \text{if } x \geq 20.$$

Thus

$$E[\text{winnings}] = \sum_{x=1}^{\infty} u(x)p_X(x) = \sum_{x=1}^{19} 2^x \left(\frac{1}{2}\right)^x + \sum_{x=20}^{\infty} 2^{20} \left(\frac{1}{2}\right)^x$$

$$= \sum_{x=1}^{19} 1 + 2^{20} \sum_{x=20}^{\infty} \left(\frac{1}{2}\right)^x = 19 + 2^{20} \frac{\left(\frac{1}{2}\right)^{20}}{1 - \frac{1}{2}} = 19 + 2^{20} \left(\frac{1}{2}\right)^{19} = 19 + 2 = \$21,$$

where we have used the formula for the sum of a geometric series with initial term $a = \left(\frac{1}{2}\right)^{20}$ and multiplicative term $r = \frac{1}{2}$. Since our expected return is $21.00, it would make long-term sense to pay $20 to play the game. (Before you go out and immediately challenge someone to allow you to pay them $20 to play the game, see Exercises 4.2 and 4.3.)

Some expected values play special roles in probability and statistical inference settings. We include them all as part of a single definition.

Definition 4.5 Let X be a random variable with p.d.f. $f_X(x)$. Some expectations that are important in a variety of settings are the following:

(a) The *k*th **moment of the distribution** (if it exists) is given by

$$\mu_k = E[X^k], k = 1, 2, \ldots .$$ (4.4)

(b) The **expected value or mean of the distribution** (if it exists) is

$$\mu = E[X].$$ (4.5)

Note that the mean is simply the first moment, μ_1, of the distribution.

(c) The **variance of the distribution** (if it exists) is

$$Var(X) = \sigma_X^2 = E\left[(X - E[X])^2\right].$$ (4.6)

The **standard deviation, σ_X, of the distribution** (if it exists) is the square root of the variance σ_X^2.

(d) For fixed $t, -\infty < t < \infty$, define

$$M_X(t) = E[e^{tX}] = \int_{-\infty}^{\infty} e^{tx} f_X(x)dx \left(\text{or} \sum_x e^{tx} f_X(x) \right).$$ (4.7)

If $M_X(t)$ exist for all $t \in (-h, h)$ for some $h > 0$, then $M_X(t)$ is called the **moment generating function (m.g.f.) for the X distribution.**

Notes
1. You are asked in Exercise 4.5 to show that

$$Var(X) = E[X^2] - \{E[X]\}^2 = \mu_2 - \mu_1^2,$$ (4.8)

when it exists. In many settings, this is the preferred way to calculate the variance for a probability distribution.
2. If the moment generating function, $M_X(t)$, exists for all $t \in (-h, h)$ and some $h > 0$ for a random variable X, then it is unique and completely determines all the properties (p.d.f., c.d.f., moments, etc.) of the associated probability distribution for X.

Example 4.12 Discrete Example
Draw a number at random from the bowl containing the integers $\{-1, -1, 2, 4, -3,$
$3, 7, 9\}$, and let X denote the number drawn. The p.d.f. for X is given by

$$f_X(x) = \frac{1}{8}, x = -3, 2, 3, 4, 7, 9$$
$$= \frac{1}{4}, x = -1$$
$$= 0, \quad \text{elsewhere.}$$

Then, we have

$$E[X] = \mu_X = \left\{ \frac{1}{8}[-3 + 2 + 3 + 4 + 7 + 9] + \frac{1}{4}(-1) \right\} = \frac{22}{8} - \frac{2}{8} = \frac{5}{2}$$

and

$$E[X^2] = \mu_2 = \left\{ \frac{1}{8}\left[(-3)^2 + (2)^2 + (3)^2 + (4)^2 + (7)^2 + (9)^2\right] + \frac{1}{4}\left[(-1)^2\right] \right\}$$
$$= \frac{168}{8} + \frac{1}{4} = \frac{85}{4},$$

so that, using expression (4.8),

$$Var(X) = \sigma_X^2 = E[X^2] - \{E[X]\}^2 = \frac{85}{4} - \left\{\frac{5}{2}\right\}^2 = \frac{85 - 25}{4} = 15.$$

Example 4.13 Rolling a Pair of Six-Sided Dice
Let X be the sum on the roll of a pair of fair six-sided dice. The p.d.f. for X is
given by

$$f_X(x) = \frac{1}{36}, \quad x = 2, 12$$
$$= \frac{2}{36}, \quad x = 3, 11$$
$$= \frac{3}{36}, \quad x = 4, 10$$
$$= \frac{4}{46}, \quad x = 5, 9$$
$$= \frac{5}{36}, \quad x = 6, 8$$
$$= \frac{6}{36}, \quad x = 7$$
$$= 0, \quad \text{elsewhere.}$$

Then, $E[X] = \mu_x$

$$= \left[\frac{1}{36}(2+12) + \frac{2}{36}(3+11) + \frac{3}{36}(4+10) + \frac{4}{36}(5+9) + \frac{5}{36}(6+8) + \frac{6}{36}(7)\right]$$

$$= \frac{14}{36}[1+2+3+4+5] + \frac{6}{36}[7] = \frac{14}{36}\left[\frac{5(6)}{2}\right] + \frac{7}{6} = \frac{42}{6} = 7.$$

Note The distribution for X in Example 4.13 is symmetric about $c = 7 = \mu_X$. Is it always true that the mean is equal to the point of symmetry for a symmetric distribution? The answer is yes, if the mean exists. We prove this result in the case of a continuous variable—just replace integrals by sums for the proof in the discrete case.

Theorem 4.1 Mean of a Symmetric Distribution
Let X be a continuous random variable with p.d.f. $f_X(x)$, and assume that its probability distribution is symmetric about the point c. Then, $\mu_X = E[X] = c$, provided the expectation exists.

Proof

$$E[X - c] = \int_{-\infty}^{\infty} (x-c)f_X(x)dx \underset{y=2c-x}{=} \int_{-\infty}^{\infty} (c-y)f_X(c+(c-y))dy$$

$$\underset{\text{symmetry}}{=} \int_{-\infty}^{\infty} (c-y)f_X(c-(c-y))dy = \int_{-\infty}^{\infty} (c-y)f_X(y)dy = E[c-X]$$

$$\Rightarrow E[X - c] = E[X] - c = E[c - X] = c - E[X] \Rightarrow 2\,E[X] = 2c \Rightarrow E[X] = c \;\blacksquare$$

Example 4.14 Mean for the Normal Distribution
Let $X \sim n(\mu, \sigma^2)$. Then

$$E[X - \mu] = \int_{-\infty}^{\infty} (x-\mu)\frac{1}{\sqrt{2\pi\sigma^2}}e^{-\frac{1}{2\sigma^2}(x-\mu)^2}dx = -\frac{-\sigma^2}{\sqrt{2\pi\sigma^2}}e^{-\frac{1}{2\sigma^2}(x-\mu)^2}\Big|_{x=-\infty}^{x=\infty}$$

$$= 0 \Rightarrow E[X] = \mu.$$

(This also follows from the fact that the $n\,(\mu, \sigma^2)$ distribution is symmetric about μ.)

Example 4.15 Moments for the Gamma Distribution
Let $X \sim Gamma\,(\alpha, \beta)$ and let $k > 0$ be arbitrary. Then

$$E[X^k] = \int_0^\infty x^k \frac{1}{\Gamma(\alpha)\,\beta^\alpha} x^{\alpha-1} e^{-\frac{x}{\beta}}\, dx$$

$$= \frac{\Gamma(\alpha+k)\beta^{\alpha+k}}{\Gamma(\alpha)\beta^\alpha} \int_0^\infty \frac{1}{\Gamma(\alpha+k)\beta^{\alpha+k}} x^{\alpha+k-1} e^{-\frac{x}{\beta}} dx$$

$$= \frac{\beta^k\,\Gamma(\alpha+k)}{\Gamma(\alpha)},$$

since the latter integrand is just the *Gamma* $(\alpha + k, \beta)$ p.d.f. and the associated integral is equal to 1. (For the moment, we will just consider this a "trick" to help us in evaluating this particular integral of interest. However, we will hopefully convince you that such a "trick" can be transformed into a useful "technique" for evaluating seemingly difficult integrals (and sums, as we shall soon see) that often arise in statistical distribution theory. While we have no problem with trying to evaluate somewhat difficult integrals directly (after all, that is one of the reasons that we learned calculus in the first place), using this valuable trick/technique will have the additional benefit of helping us become very comfortable with the form and properties of the most common continuous and discrete probability distributions that we encounter in statistics.)

Special cases:

$$k = 1 \Rightarrow E[X] = \mu_X = \frac{\beta\Gamma(\alpha+1)}{\Gamma(\alpha)} = \frac{\beta\alpha\Gamma(\alpha)}{\Gamma(\alpha)} = \alpha\beta$$

$$k = 2 \Rightarrow E[X^2] = \frac{\beta^2\,\Gamma(\alpha+2)}{\Gamma(\alpha)} = \frac{\beta^2(\alpha+1)\alpha\Gamma(\alpha)}{\Gamma(\alpha)} = \alpha(\alpha+1)\beta^2$$

Thus,

$$Var(X) = \sigma_X^2 = E[X^2] - \{E[X]\}^2$$

$$= \alpha(\alpha+1)\beta^2 - (\alpha\beta)^2 = \alpha^2\beta^2 + \alpha\beta^2 - \alpha^2\beta^2 = \alpha\beta^2.$$

Example 4.16 Mean and Variance for the Binomial Distribution
Let $X \sim Binom\ (n, p)$. The p.d.f. for X is given by

$$f_X(x) = \binom{n}{x} p^x (1-p)^{n-x} I_{\{0,1,\ldots,n\}}(x)$$

$$\Rightarrow E[X] = \mu_X$$

$$= \sum_{x=0}^{n} x \frac{n!}{x!(n-x)!} p^x (1-p)^{n-x} = \sum_{x=1}^{n} \frac{n!}{(x-1)!(n-x)!} p^x (1-p)^{n-x}$$

$$= \sum_{y=x-1}^{n-1} \sum_{y=0}^{n-1} \frac{n!}{y!(n-y-1)!} p^{y+1} (1-p)^{n-y-1} = np \sum_{y=0}^{n-1} \frac{(n-1)!}{y!(n-1-y)!} p^y (1-p)^{n-1-y}$$

$$= np \sum_{y=0}^{n-1} \binom{n-1}{y} p^y (1-p)^{(n-1)-y} = np,$$

since the last sum is just the complete sum of the *Binom* (n-1, p) p.d.f. and, thus, equal to 1. (Is it a "technique" yet?)
Similarly,

$$E[X(X-1)] = \sum_{x=0}^{n} x(x-1) \frac{n!}{x!(n-x)!} p^x (1-p)^{n-x} = \sum_{x=2}^{n} \frac{n!}{(x-2)!(n-x)!} p^x (1-p)^{n-x}$$

$$= \sum_{y=x-2}^{n-2} \sum_{y=0}^{n-2} \frac{n!}{y!(n-y-2)!} p^{y+2} (1-p)^{n-y-2} = n(n-1)p^2 \sum_{y=0}^{n-2} \frac{(n-2)!}{y!(n-2-y)!} p^y (1-p)^{n-2-y}$$

$$= n(n-1)p^2 \sum_{y=0}^{n-2} \binom{n-2}{y} p^y (1-p)^{(n-2)-y} = n(n-1)p^2,$$

since the last sum is just the sum of the *Binom* (n-2, p) p.d.f. and, thus, equal to 1. (Are you there yet with the "trick" to "technique"?)

It follows that

$$E[X^2] = E[X(X-1)] + E[X] = n(n-1)p^2 + np$$
$$\Rightarrow Var(X) = \sigma_X^2 = E[X^2] - \{E[X]\}^2 = n(n-1)p^2 + np - (np)^2$$
$$= -np^2 + np = np(1-p).$$

Thus, if $X \sim Binom$ (n, p), then $\mu_X = np$ and $\sigma_X^2 = np(1-p)$.

Example 4.17 Moment Generating Function for the Poisson Distribution
Let $X \sim Poisson$ (λ), with $\lambda > 0$. Then X has p.d.f.

$$f_X(x) = \frac{\lambda^x e^{-\lambda}}{x!} I_{\{0,1,2,\ldots\}}(x), \quad \text{with } \lambda > 0.$$

$\Rightarrow$ the moment generating function (m.g.f.) for X is

$$M_X(t) = E[e^{tX}] = \sum_{x=0}^{\infty} e^{tx} \frac{\lambda^x e^{-\lambda}}{x!}$$

$$= e^{-\lambda} \sum_{x=0}^{\infty} \frac{(\lambda e^t)^x}{x!} = e^{-\lambda} e^{\lambda e^t} = e^{\lambda(e^t - 1)}, \quad -\infty < t < \infty.$$

Think About It How do we know that $\sum_{x=0}^{\infty} \frac{(\lambda e^t)^x}{x!} = e^{\lambda e^t}$?

Example 4.18 Moment Generating Function for the Binomial Distribution
Let $X \sim Binom\,(n, p)$, with $0 \leq p \leq 1$. Then X has p.d.f.

$$f_X(x) = \binom{n}{x} p^x (1-p)^{n-x} I_{\{0,1,\dots,n\}}(x)$$

and its moment generating function (m.g.f.) is given by

$$M_X(t) = E[e^{tX}] = \sum_{x=0}^{n} e^{tx} \binom{n}{x} p^x (1-p)^{n-x}$$

$$= \sum_{x=0}^{n} \binom{n}{x} (pe^t)^x (1-p)^{n-x}$$

$$= [(1-p) + pe^t]^n, \quad -\infty < t < \infty.$$

Think About It How do we know that $\sum_{x=0}^{n} \binom{n}{x} (pe^t)^x (1-p)^{n-x} = [(1-p) + pe^t]^n$, for $-\infty < t < \infty$?

Example 4.19 Moment Generating Function for the Geometric Distribution
Let $X \sim Geom\,(p)$, with $0 < p \leq 1$. Then X has p.d.f.

$$f_X(x) = p(1-p)^{x-1} I_{\{1,2,3,\dots\}}(x)$$

and its m.g.f. is given by

$$M_X(t) = E[e^{tX}] = \sum_{x=1}^{\infty} e^{tx} p(1-p)^{x-1}$$

$$= \frac{p}{1-p} \sum_{x=1}^{\infty} [(1-p)e^t]^x.$$

Note that this sum is just a geometric series with initial term $a = (1-p)e^t$ and multiplicative term $r = (1-p)e^t$ as well and it converges as long as $r = (1-p)e^t < 1$ or, equivalently, $-\infty < t < -\ln(1-p)$. Under this condition, the geometric series converges to

$$\frac{a}{1-r} = \frac{(1-p)e^t}{1-(1-p)e^t},$$

and it follows that

$$M_X(t) = \frac{p}{(1-p)} \frac{(1-p)e^t}{1-(1-p)e^t} = \frac{pe^t}{1-(1-p)e^t}, \quad -\infty < t < -\ln(1-p).$$

Think About It One of the requirements for the moment generating function to exist is that the associated sum (or integral in the case of continuous variables) must exist in a symmetric interval centered at 0. Is this satisfied in Example 4.19?

Example 4.20 Moment Generating Function for the Roll of a Fair Die
Let X represent the outcome of the roll of a single fair die. Then X has p.d.f.

$$f_X(x) = \frac{1}{6}I_{\{1,2,3,4,5,6\}}(x)$$

and its m.g.f. is given by

$$M_X(t) = E\left[e^{tX}\right] = \sum_{x=1}^{6} e^{tx}\frac{1}{6} = \frac{1}{6}e^t + \frac{1}{6}e^{2t} + \cdots + \frac{1}{6}e^{6t}, \quad -\infty < t < \infty.$$

Example 4.21 Moment Generating Function for the Gamma Distribution
Let $X \sim Gamma\,(\alpha, \beta)$, with $\alpha > 0$ and $\beta > 0$. Then X has p.d.f.

$$f_X(x) = \frac{1}{\Gamma(\alpha)\beta^\alpha}x^{\alpha-1}e^{-\frac{x}{\beta}}I_{(0,\infty)}(x)$$

and its m.g.f. is given by

$$M_X(t) = E[e^{tX}] = \int_0^\infty e^{tx}\frac{1}{\Gamma(\alpha)\beta^\alpha}x^{\alpha-1}e^{-\frac{x}{\beta}}dx$$

$$= \int_0^\infty \frac{1}{\Gamma(\alpha)\beta^\alpha}x^{\alpha-1}e^{-x\left(\frac{1}{\beta}-t\right)}dx = \int_0^\infty \frac{1}{\Gamma(\alpha)\beta^\alpha}x^{\alpha-1}e^{-x/\left(\frac{\beta}{1-\beta t}\right)}dx$$

$$= \frac{1}{\beta^\alpha}\left(\frac{\beta}{1-\beta t}\right)^\alpha \int_0^\infty \frac{1}{\Gamma(\alpha)\left[\frac{\beta}{1-\beta t}\right]^\alpha}x^{\alpha-1}e^{-x/\left(\frac{\beta}{1-\beta t}\right)}dx$$

But the integrand in the latter integral is simply a gamma p.d.f. with parameters α and $\frac{\beta}{1-\beta t}$, provided that $\frac{\beta}{1-\beta t} > 0 \Leftrightarrow 1-\beta t > 0 \Leftrightarrow t < \frac{1}{\beta}$. Thus, when $t < \frac{1}{\beta}$, the latter integral is simply equal to 1, and it follows that

$$M_X(t) = \frac{\left(\frac{\beta}{1-\beta t}\right)^\alpha}{\beta^\alpha} = (1-\beta t)^{-\alpha}, \quad -\infty < t < \frac{1}{\beta}.$$

Think About It Okay, have you turned the "trick" into a "technique" by now? Incidentally, does the associated integral for the m.g.f. for the gamma distribution exist in the required symmetric interval centered at 0?

Example 4.22 Moment Generating Function for the Normal Distribution
Let $X \sim n\,(\mu, \sigma^2)$. Then X has p.d.f.

$$f_X(x) = \frac{1}{\sqrt{2\pi\sigma^2}}\, e^{-\frac{1}{2\sigma^2}(x-\mu)^2}\, I_{(-\infty,\infty)}(x)$$

and its m.g.f. is given by

$$
\begin{aligned}
M_X(t) = E\big[e^{tX}\big] &= \int_{-\infty}^{\infty} e^{tx}\, \frac{1}{\sqrt{2\pi\sigma^2}}\, e^{-\frac{1}{2\sigma^2}(x-\mu)^2}\, dx \\
&= \int_{-\infty}^{\infty} \frac{1}{\sqrt{2\pi\sigma^2}}\, e^{-\frac{1}{2\sigma^2}\left[x^2 - 2\mu x + \mu^2 - 2\sigma^2 tx\right]}\, dx \\
&\underset{\text{completing the square}}{=} \int_{-\infty}^{\infty} \frac{1}{\sqrt{2\pi\sigma^2}}\, e^{-\frac{1}{2\sigma^2}\left[x^2 - 2x\left(\mu+\sigma^2 t\right) + \mu^2 \pm \left(\mu+\sigma^2 t\right)^2\right]}\, dx \\
&= e^{-\frac{\mu^2}{2\sigma^2}} e^{\frac{\left(\mu+\sigma^2 t\right)^2}{2\sigma^2}} \int_{-\infty}^{\infty} \frac{1}{\sqrt{2\pi\sigma^2}}\, e^{-\frac{1}{2\sigma^2}\left[x - \left(\mu+\sigma^2 t\right)\right]^2}\, dx.
\end{aligned}
$$

But the latter integrand is just the p.d.f. for a normal distribution with mean $\mu + \sigma^2 t$ and variance σ^2. (Technique!!) Hence the integral is simply equal to 1, and it follows that

$$M_X(t) = e^{\frac{-\mu^2 + \mu^2 + 2\mu\sigma^2 t + \sigma^4 t^2}{2\sigma^2}} = e^{\mu t + \frac{\sigma^2 t^2}{2}}, \quad -\infty < t < \infty.$$

The moment generating function has a number of important uses in statistical distribution theory, as we shall see throughout the text. The first of these is that it naturally (no surprise, from its name!) can be used to generate the moments of a probability distribution, most importantly the mean and variance for the distribution.

Theorem 4.2 Generating Moments from the Moment Generating Function
Let X be a random variable with p.d.f. $f_X(x)$ and m.g.f. $M_X(t) = E[e^{tX}]$. Let k be a positive integer such that $E[X^k]$ exists. Then under the assumption that we can move derivatives inside either the integral (for the continuous case, as illustrated here) or the sum (for the discrete case), we have the following result:

$$\left.\frac{d^k M_X(t)}{dt^k}\right|_{t=0} = E\big[X^k\big]. \tag{4.9}$$

Proof Consider first the case $k = 1$. Then under the assumption of being able to move the derivative inside the integral, we have

$$\frac{dM_X(t)}{dt} = \frac{d}{dt}\left[E[e^{tX}]\right] = \frac{d}{dt}\left[\int_{-\infty}^{\infty} e^{tx} f_X(x)dx\right] \text{ "} = \text{" } \int_{-\infty}^{\infty} \frac{d}{dt} e^{tx} f_X(x)dx = \int_{-\infty}^{\infty} x e^{tx} f_X(x)dx$$

$$\Rightarrow \frac{dM_X(t)}{dt}\bigg|_{t=0} = \int_{-\infty}^{\infty} x f_X(x)dx = \mu_X = E[X].$$

In general, for arbitrary positive integer k such that $E[X^k]$ exists, we repeat this derivative process k times before evaluating the result at $t = 0$, yielding

$$E[X^k] = \int_{-\infty}^{\infty} x^k f_X(x)dx = \int_{-\infty}^{\infty} x^k e^{tx}\big|_{t=0} f_X(x)dx = \int_{-\infty}^{\infty} \frac{d^k e^{tx}}{dt^k}\bigg|_{t=0} f_X(x)dx$$

$$\text{"} = \text{" } \frac{d^k}{dt^k} \int_{-\infty}^{\infty} e^{tx} f_X(x)dx\big|_{t=0} = \frac{d^k M_X(t)}{dt^k}\bigg|_{t=0} = \mu_k,$$

the kth moment of the X distribution ∎

Note $Var(X) = \sigma_X^2 = E[X^2] - \mu_X^2 = \frac{d^2}{dt^2} M_X(t)\big|_{t=0} - \left\{\frac{d}{dt} M_X(t)\big|_{t=0}\right\}^2.$

Definition 4.6 Let X be a random variable with moment generating function $M_X(t)$. The **cumulant generating function** for X is given by $\Psi_X(t) = \ln M_X(t)$.

Corollary 4.1 Generating the Mean and Variance from the Cumulant Generating Function

Let X be a random variable with mean μ_X, variance σ_X^2, m.g.f. $M_X(t)$, and associated cumulant generating function $\Psi_X(t)$. Then under the conditions of Theorem 4.2, it follows that

$$\frac{d}{dt} \Psi_X(t)\big|_{t=0} = \mu_X$$

and

$$\frac{d^2}{dt^2} \Psi_X(t)\big|_{t=0} = \sigma_X^2.$$

Proof Exercise 4.8 ∎

Example 4.23 Poisson Distribution

Let $X \sim Poisson(\lambda)$. Then

$$\Psi_X(t) = \ln M_X(t) = \ln\left[e^{\lambda(e^t - 1)}\right] = \lambda(e^t - 1)$$

$$\Rightarrow \quad \mu_X = \frac{d}{dt}\Psi_X(t)|_{t=0} = \lambda e^t|_{t=0} = \lambda$$

and

$$\sigma_X^2 = \frac{d^2}{dt^2}\Psi_X(t)|_{t=0} = \frac{d}{dt}\lambda e^t|_{t=0} = \lambda e^t|_{t=0} = \lambda.$$

Example 4.24 Normal Distribution

Let $X \sim n(\theta, \tau^2)$. Then,

$$\Psi_X(t) = \ln M_X(t) = \ln\left[e^{\theta t + \frac{\tau^2 t^2}{2}}\right] = \theta t + \frac{\tau^2 t^2}{2}$$

$$\Rightarrow \quad \mu_X = \frac{d}{dt}\Psi_X(t)|_{t=0} = \theta + \frac{2\tau^2 t}{2}\bigg|_{t=0} = \theta$$

and

$$\sigma_X^2 = \frac{d^2}{dt^2}\Psi_X(t)|_{t=0} = \frac{d}{dt}\left[\theta + \tau^2 t\right]\big|_{t=0} = \tau^2|_{t=0} = \tau^2.$$

Example 4.25 Gamma Distribution

Let $X \sim Gamma(\alpha, \beta)$, with $\alpha > 0$ and $\beta > 0$. Then

$$M_X(t) = (1 - \beta t)^{-\alpha}, \text{ for } t < 1/\beta$$

$$\Rightarrow \quad \mu_X = \frac{d}{dt}M_X(t)|_{t=0} = -\alpha(1 - \beta t)^{-\alpha - 1}(-\beta)|_{t=0} = \alpha\beta$$

and

$$E[X^2] = \frac{d^2}{dt^2}M_X(t)|_{t=0} = (-\alpha)(-\beta)(-\alpha - 1)(1 - \beta t)^{-\alpha - 2}(-\beta)|_{t=0} = \alpha\beta^2(1 + \alpha),$$

so that

$$\sigma_X^2 = E[X^2] - \{\mu_X\}^2 = \alpha\beta^2(1 + \alpha) - (\alpha\beta)^2 = \alpha\beta^2.$$

Example 4.26 Binomial Distribution

Let $X \sim Binom\,(n, p)$. Then

$$M_X(t) = \{(1 - p) + pe^t\}^n$$

so that

$$\mu_X = \frac{d}{dt}\{(1-p) + pe^t\}^n\Big|_{t=0} = n\{(1 - p) + pe^t\}^{n-1} pe^t\big|_{t=0}$$

$$= npe^0\{(1-p) + pe^0\}^{n-1} = np$$

and

$$E[X^2] = \frac{d}{dt}\left[npe^t\{(1-p) + pe^t\}^{n-1}\right]\Big|_{t=0}$$

$$= npe^t\{(1-p) + pe^t\}^{n-1}\Big|_{t=0} + n(n-1)pe^t\{(1-p) + pe^t\}^{n-2} pe^t\big|_{t=0}$$

$$= np + n(n-1)p^2$$

$$\Rightarrow \quad \sigma_X^2 = E[X^2] - \mu_X^2 = np + n(n-1)p^2 - (np)^2 = np - np^2 = np(1-p).$$

Example 4.27 Roll of a Fair Die

Let X be the outcome for the roll of a single fair die. Then

$$M_X(t) = \sum_{x=1}^{6} xe^{tx}$$

and

$$\mu_X = \frac{d}{dt} M_X(t)\Big|_{t=0} = \sum_{x=1}^{6} \frac{xe^{tx}}{6}\Big|_{t=0} = \sum_{x=1}^{6} \frac{x}{6} = \frac{6(7)}{2(6)} = \frac{7}{2}.$$

Similarly,

$$E[X^2] = \frac{d^2}{dt^2} M_X(t)\Big|_{t=0} = \sum_{x=1}^{6} \frac{x^2 e^{tx}}{6}\Big|_{t=0} = \sum_{x=1}^{6} \frac{x^2}{6} = \frac{6(7)(2(6)+1)}{6(6)} = \frac{91}{6},$$

so that

$$\sigma_X^2 = \frac{91}{6} - \left(\frac{7}{2}\right)^2 = \frac{182 - 147}{12} = \frac{35}{12}.$$

Think About It Does the m.g.f. approach to obtaining $E[X] = \mu_X$ and $E[X^2]$ actually have any advantage in Example 4.27 over obtaining these moments directly from their definitions?

4.5   Chebyshev's Inequality

The mean, μ_X, provides a natural measure of the center for the probability distribution of a random variable X, and the variance, σ_X^2, provides a natural measure of the variability associated with the probability distribution. However, these two measures also interact to put additional constraints on the probability distribution. One such constraint is illustrated by the well-known Chebyshev's inequality.

Theorem 4.3 Chebyshev's Inequality
Let X be a random variable with p.d.f. $f_X(x)$, mean $\mu_X = E[X]$, and variance $\sigma_X^2 = Var(X) < \infty$. Then for any positive constant $m > 0$, we have

$$P(|X - \mu_X| \geq m\sigma_X) = P\left(\left|\frac{X - \mu_X}{\sigma_X}\right| \geq m\right) \leq \frac{1}{m^2}. \qquad (4.10)$$

Note: This is equivalent to

$$P(|X - \mu_X| < m\sigma_X) = P\left(\left|\frac{X - \mu_X}{\sigma_X}\right| < m\right) \geq 1 - \frac{1}{m^2}.$$

Proof We provide the proof in the continuous case, with the proof for the discrete case obtained by simply replacing the integrals by corresponding summations. We have

$$\sigma_X^2 = E\left[(X - \mu_X)^2\right] = \int_{-\infty}^{\infty} (x - \mu_X)^2 f_X(x)dx$$

$$= \int_A (x - \mu_X)^2 f_X(x)dx + \int_{A^C} (x - \mu_X)^2 f_X(x)dx,$$

where $A = \{x : |x - \mu_X| \geq m\sigma_X\}$. Since $\int_{A^C} (x - \mu_X)^2 f_X(x)dx \geq 0$, it follows from the definition of the set A that

$$\sigma_X^2 \geq \int_A (x - \mu_X)^2 f_X(x)dx \geq \int_A m^2 \sigma_X^2 f_X(x)dx \geq m^2 \sigma_X^2 \int_A f_X(x)dx = m^2 \sigma_X^2 P(A)$$

$$= m^2 \sigma_X^2 P(|X - \mu_X| \geq m\sigma_X)$$

$$\Rightarrow \quad P(|X - \mu_X| \geq m\sigma_X) \leq \frac{\sigma_X^2}{m^2 \sigma_X^2} = \frac{1}{m^2}. \quad \blacksquare$$

In particular, Chebyshev's inequality provides the following specific restrictions on the probability distribution for a random variable X with mean μ_X and variance σ_X^2:

1. $P(|X - \mu_X| < 2\sigma_X) \geq 1 - \frac{1}{2^2} = .750$
2. $P(|X - \mu_X| < 3\sigma_X) \geq 1 - \frac{1}{3^2} = \frac{8}{9} = .889$.

Thus, the probability is relatively small that we observe values of a random variable X too many standard deviations away from its mean, no matter what the probability distribution for X looks like.

Examples 4.28 Roll of a Fair Die
Let X be the outcome for the roll of a fair six-sided die. Then we have seen (Example 4.27) that $\mu_X = 3.5$ and $\sigma_X^2 = \frac{35}{12}$. From Chebyshev's inequality with $m = 2$, we see that

$$P\left(|X - 3.5| < 2\sqrt{\frac{35}{12}}\right) = P(|X - 3.5| < 3.416) \geq .75.$$

The actual probability is

$$P(|X - 3.5| < 3.416) = P(0 < X < 6.9) = 1.$$

Example 4.29 Exponential Distribution
Let $X \sim Exp\,(1)$ with p.d.f.

$$f_X(x) = e^{-x} I_{(0,\infty)}(x).$$

Then $\mu_X = 1$ and $\sigma_X^2 = 1$, so that Chebyshev's inequality with $m = 2$ states that

$$P(|X - 1| < 2(1)) = P(-1 < X < 3) = P(0 < X < 3) \geq .75.$$

The actual probability is

$$P(0 < X < 3) = \int_0^3 e^{-x}dx = -e^{-x}\bigg|_{x=0}^{x=3} = 1 - e^{-3} = 1 - .050 = .95.$$

Think About It In these two examples, Chebyshev's inequality does not provide a particularly sharp lower bound for the probability that a random variable deviates from its mean by more than two standard deviations. Does that make you think that we should be able to improve on the inequality (i.e., make it sharper)?

The following example shows, in fact, that the Chebyshev bound for all distributions cannot be made any sharper without imposing further restrictions on the probability distribution.

Example 4.30 Example Where the Chebyshev Bound Is Attained
Let X be a discrete random variable with p.d.f.

$$f_X(x) = \frac{1}{8}, \quad x = -1, +1$$
$$= \frac{6}{8}, \quad x = 0$$
$$= 0, \quad \text{elsewhere.}$$

Then, we have

$$E[X] = \mu_X = (-1)\frac{1}{8} + (1)\frac{1}{8} + (0)\frac{6}{8} = 0$$

and

$$\sigma_X^2 = E[X^2] = (-1)^2\left(\frac{1}{8}\right) + (1)^2\left(\frac{1}{8}\right) + (0)^2\frac{6}{8} = \frac{1}{4}.$$

It follows from Chebyshev's inequality with $m = 2$ that

$$P(|X - \mu_X| < 2\sigma_X) = P\left(|X - 0| < 2\sqrt{\frac{1}{4}}\right) = P(|X| < 1) \ge .75.$$

On the other hand, the exact probability is $P(|X| < 1) = P(X = 0) = 0.75$ as well, so that Chebyshev's bound is actually achieved (i.e., it cannot be improved) for this distribution. (One could construct a continuous distribution that also achieves the bound.)

4.6   Exercises

4.1. Suppose X is a continuous random variable with c.d.f. $F_X(x)$ that satisfies the condition in (4.2) for all x and some c. Show that the probability distribution for X also satisfies the condition in (4.1) for this point c and all x, so that the probability distribution is symmetric about the point c.

4.2. Consider the setting of Example 4.11. Just because the expected winnings are positive when you pay \$20 to play the game, there are other things to consider when making a decision to do so. In particular, how large does X have to be before we will see a profit on an individual game? What is the probability of an X at least this large for any given game? What does this imply about our ability to show a profit in the long run from playing this game? (Of course, we could simply be *really lucky* and get tails on the first 19 tosses of the coin before getting the first head on the 20th toss, thereby putting away a tidy profit of $\$2^{20} = \$1,048,576$, less our initial \$20 fee to play! What is the probability of this happening?)

4.3. Consider again the setting of Example 4.11. Write down an expression for the probability that you will have had at least one opportunity to be ahead (i.e., make money) if you are willing to play the game up to a fixed maximum number, say k, of times; that is, you have $\$20\,k$ available to gamble with—and you are willing to risk losing that much, but no more, before winning overall. Try it for $k = 1$, then $k = 2$, etc. Is there any hope of selecting an optimal k for playing this game?

4.4. Let's return one last time to the setting of Example 4.11, but I am feeling generous. I will allow you to charge me whatever you want to play this game if you will simply tweak the payout formula to the following:

$$u(x) = 2^x \text{ for all } x = 1, 2, 3, \ldots$$

Should you agree to play the game with me? If so, what should you charge me to play the game to make it fair to you? (This is called the St. Petersburg Paradox.) [Note: Keep in the mind the caution discussed in Exercise 4.2.]

4.5. Let X be a random variable. Show that

$$Var(X) = E\left[X^2\right] - \{E[X]\}^2 = \mu_2 - \mu_1^2,$$

provided it exists.

4.6. Let X be the sum of the roll of two fair dice. Find $\sigma_X^2 = Var(X)$. (See Example 4.13.)

4.7. Using the same approach that we used with the binomial distribution in Example 4.16, find $E[X]$, $E[X(X-1)]$, and $Var(X)$ for the Poisson (λ) distribution.

4.8. Prove Corollary 4.1.

4.9. Let X be a random variable with mean μ_X and finite variance σ_X^2, and let a and b be arbitrary constants. Show that $E[aX + b] = a\,\mu_X + b$ and $Var(aX + b) = a^2\sigma_X^2$.

4.10. Let X be a continuous random variable with p.d.f.

$$f_X(x) = \frac{1}{2}\left[e^{-x} + xe^{-x}\right]I_{(0,\infty)}(x).$$

Find the moment generating function for this distribution and use it to obtain $E[X]$ and $Var(X)$.

4.11. Let X be a continuous random variable with p.d.f.

$$f_X(x) = \frac{1}{x}I_{(1,e)}(x).$$

(a) Find $E[X]$ and $Var(X)$.
(b) Evaluate $E\left[\frac{\ln(X)}{X}\right]$.
(c) Find the c.d.f. for X.
(d) Use the c.d.f. to evaluate $P(1 < X < 2)$.
(e) Obtain the median of the X distribution.
(f) Find $E[\ln^k(X)]$ for an arbitrary positive integer k.

4.12. Let X be a continuous random variable with p.d.f.

$$f_X(x) = 6x(1 - x)I_{(0,1)}(x).$$

(a) Find the c.d.f. for X.
(b) Evaluate $P(.5 < X < .75)$.
(c) Find $Var(X)$.
(d) Find $E\left[\frac{\ln(X)}{X(1-X)}\right]$.

4.13. Let X be a continuous random variable with

$$f_X(x) = \theta e^{\theta x}I_{(-\infty,0)}(x),$$

where $\theta > 0$.

(a) Find the moment generating function for X.
(b) Use the moment generating function to obtain $E[X]$ and $Var(X)$.

4.14. Let X have a Poisson distribution with parameter $\lambda > 0$. Under what conditions on λ does $E[X!]$ exist? Find a closed form expression for $E[X!]$ under those conditions.

4.15. Let X be a continuous random variable with p.d.f.

$$f_X(x) = 2xI_{(0,1)}(x).$$

(a) Find the c.d.f. for X.
(b) Find $E[X]$ and $Var(X)$ directly.
(c) Derive the moment generating function for X.
(d) Use the moment generating function from part (c) to verify your results in part (b).
(e) Find the value of the median for the X distribution.

4.16. Let X be a discrete random variable with p.d.f. $f_X(x)$ given by the following table:

x	1	2	3	6	7	8	20
$f_X(x)$	.10	.15	.10	.15	.15	.05	.30

(a) Find the c.d.f. for X.
(b) Directly evaluate $E[X]$ and $Var(X)$.
(c) Derive the moment generating function for X and use it to verify the results in (b).
(d) Find the value of the median(s) for the X distribution.

4.17. Let X be a continuous random variable with p.d.f.

$$f_X(x) = [.6e^{-x} + .4xe^{-x}]I_{(0,\infty)}(x).$$

Find $E[X]$ and $Var(X)$.

4.18. Let $X \sim Geom\,(p)$.

(a) Obtain a closed form expression for $P(X > x)$, $x \in \{1, 2, 3, \ldots\}$.
(b) Show that the probability distribution for X satisfies the property
$P(X > s + t) = P(X > s)\,P(X > t)$ for every pair of positive integers (s, t).

Think About It Why do you think this is known as the *memoryless property*?

4.19. Let X be a continuous random variable with c.d.f.

$$F_X(x) = \frac{1}{1 + e^{-x}}\,I_{(-\infty,\infty)}(x).$$

(a) Find the value of the unique median for the X distribution.
(b) Obtain the p.d.f. for X.
(c) Evaluate $P(0 < X < 2)$.
(d) Show that the X distribution is symmetric about its median.
(e) Find $E[(1 + e^{-X})^{-1}]$.

4.20. Let X be a continuous random variable with p.d.f.

$$f_X(x) = \frac{1}{\Gamma(5)2^5} x^4 e^{-\frac{x}{2}} I_{(0,\infty)}(x).$$

Let $Y = 1/X$. Find $E[Y]$ and $Var(Y)$.

4.21. Let X be a continuous random variable with moment generating function

$$M_X(t) = \frac{e^{\theta t}}{1 - \eta^2 t^2} I_{(-\frac{1}{\eta}, \frac{1}{\eta})}(t),$$

where $0 < \eta < \infty$ and $-\infty < \theta < \infty$. Find $E[X]$ and $Var(X)$.

4.22. Below are two functions defined on the real line. One of them is a probability density function for some random variable, and one of them is a cumulative distribution function for some random variable.

$$h_1(t) = \left[\frac{1}{4} e^{-\frac{t}{2}} + \frac{t}{2} e^{-t}\right] I_{(0,\infty)}(t)$$

$$h_2(t) = \left[1 - e^{-\frac{t}{2}} - \frac{t}{2} e^{-\frac{t}{2}}\right] I_{(0,\infty)}(t).$$

(a) Identify which of the two functions is the cumulative distribution function. Justify your answer.
(b) Let Y be the random variable with the cumulative distribution function identified in part (a). Evaluate and simplify the expression for $P(1 < Y < 2)$.
(c) Obtain the probability density function for the random variable defined in part (b).

4.23. Let $X \sim Geom\ (p)$. Find the moment generating function for X, and use it to find $E[X]$ and $Var(X)$.

4.24. Let X be a continuous random variable with p.d.f.

$$f_X(x) = \frac{x^2 e^{-\left(\frac{x}{4}\right)}}{128} I_{(0,\infty)}(x).$$

(a) Find the form of the c.d.f. for X.
(b) Evaluate $P(1.5 < X < 2)$.
(c) Find $E[X]$ and $Var(X)$.

4.25. Let $M_X(t)$ be the moment generating function for a random variable X, and assume that all derivatives of $M_X(t)$ exist at $t = 0$.

(a) Find the form of the Taylor series expansion of $M_X(t)$ about $t = 0$.
(b) The first four terms of the Taylor series expansion about $t = 0$ of the moment generating function for the random variable Y are given by

$$M_X(t) = 1 + 12t + 90t^2 + 540t^3.$$

What is the variance for the random variable Y?

4.26. Consider a random experiment in which either event A occurs or it does not. Let p denote the probability that the experiment results in event A and let X denote the number of independent trials of the experiment necessary to achieve the event A for the first time.

(a) Evaluate the probability that the first event A occurs on an even numbered trial. Obtain a closed form expression for this probability as a function of p.
(b) Given that the event A first occurred on an even numbered trial, what is the probability that it took at least six trials to achieve it? Again, obtain a closed form expression as a function of p.

4.27. Let X be a continuous random variable with p.d.f.

$$f_X(x) = \left[\frac{1}{4}e^{-\frac{x}{2}} + \frac{1}{2}xe^{-x} \right] I_{(0,\infty)}(x).$$

(a) Find the form of the c.d.f. for X.
(b) Use the c.d.f. to evaluate $P(2 < X < 3)$.

4.28. Let X be a discrete random variable with c.d.f.

$$\begin{aligned}
F_X(x) &= 0, & -\infty < x < 2, \\
&= .3, & 2 \le x < 3, \\
&= .3, & 3 \le x < 5, \\
&= .45, & 5 \le x < 7.3, \\
&= .83, & 7.3 \le x < 9, \\
&= .9, & 9 \le x < 100, \\
&= 1, & 100 \le x < \infty.
\end{aligned}$$

(a) Evaluate $P(3.5 < X \le 7.3)$.
(b) Find the form of the probability density function for X.

4.29. Let X be a continuous random variable with p.d.f.

$$f_X(x) = \frac{1}{2}e^{-|x|}I_{(-\infty,\infty)}(x).$$

(a) Find the c.d.f. for X.

(b) Find the moment generating function for X.

Note This distribution is the standard form of what is known as the *double exponential* or *Laplace* distribution.

4.30. Below are three functions defined on the reals. One of them is a probability density function for some random variable, one of them is a moment generating function for some random variable, and one of them is a cumulative distribution function for some random variable. Identify which is which.

$$h_1(t) = \left[\frac{5}{6} + \frac{1}{12}e^{-\frac{t}{2}} + \frac{1}{12}e^{\frac{t}{2}} \right] I_{(-\infty,\infty)}(t)$$

$$h_2(t) = \left[\frac{1}{4}e^{-\frac{t}{2}} + \frac{t}{2}e^{-t} \right] I_{[0,\infty)}(t)$$

$$h_3(t) = \left[1 - e^{-\frac{t}{2}} - \frac{t}{2}e^{-\frac{t}{2}} \right] I_{[0,\infty)}(t).$$

4.31. Let X denote the random variable associated with the moment generating function in Exercise 4.30.

(a) Find the associated p.d.f. for X.

(b) Evaluate $E[X]$ and $Var(X)$.

4.32. Let Y denote the random variable associated with the cumulative distribution function in Exercise 4.30.

(a) Evaluate and simplify an expression for $P(1 < Y < 2)$.

(b) Obtain the probability distribution function for Y.

4.33. Let Z denote the random variable associated with the probability density function in Exercise 4.30.

(a) Find the cumulative distribution function for Z.

(b) Find $E\left[e^{\frac{Z}{4}}\right]$.

4.34. Consider a random experiment in which either event A occurs or it does not. Let p denote the probability that the experiment results in event A, and let X denote the number of independent trials of the experiment necessary to achieve the event A for the first time.

(a) Evaluate the probability that the first event A occurs on an even numbered trial. Obtain a closed form expression for this probability as a function of p.

(b) Given that the event A first occurred on an even numbered trial, what is the probability that it took at least five trials to achieve it? Again, obtain a closed form expression as a function of p.

(c) Compare your answers to parts (a) and (b) with those obtained in Exercise 4.26.

4.35. Let $X_n \sim$ Poisson $(n\lambda)$.

(a) What is the moment generating function for X_n?

(b) Find an expression for

$$Q_n(t) = E\left[e^{t\left\{\frac{X_n - n\lambda}{\sqrt{n\lambda}}\right\}}\right].$$

(c) Evaluate $\lim_{n \to \infty} Q_n(t)$. [Hint: Consider a Taylor series expansion of $Q_n(t)$.]

4.36. Let X be a random variable with moment generating function $M_X(t)$, mean μ_X, and variance σ_X^2. Let c be a constant and set $W = cX$.

(a) Find the moment generating function for W in terms of c and $M_X(t)$.

(b) Use the moment generating function for W to show that $E[W] = c\mu_X$ and $Var(W) = c^2 \sigma_X^2$.

4.37. Let X be a continuous random variable with p.d.f. given by

$$f_X(x) = 2xe^{-x^2} I_{(0,\infty)}(x).$$

(a) Find the c.d.f. for X.

(b) Evaluate $P(2 < X < 5)$.

(c) Find $E[X]$ and $Var(X)$.

4.38. Consider the following two experiments involving independent Bernoulli trials with common probability of success p.

Experiment 1 Conduct a fixed number of trials, n, and let X be the number of successes in the n trials.

Experiment 2 Conduct Bernoulli trials until we observe the first success. Let Y be the required number of trials.

(a) Identify the probability distribution for X and write down its p.d.f.

(b) Identify the probability distribution for Y and write down its p.d.f.

(c) Compute $P(X = 1)$ and $P(Y = n)$, and show that

$$R = \frac{P(X = 1)}{P(Y = n)}$$

does not depend on the probability of success, p. Provide an interpretation for the value of R.

(d) Let $t \in \{1, 2, \ldots, n\}$ and consider the following third experiment:

Experiment 3   Conduct Bernoulli trials until we obtain t successes, and let W be the required number of trials.

Show that

$$Q = \frac{P(X = t)}{P(W = n)}$$

is a constant that does not depend on the value of p.

4.39. Let X be a continuous random variable with p.d.f.

$$f_X(x) = \frac{1}{\lambda} e^{-\frac{(x-\alpha)}{\lambda}} I_{(\alpha,\infty)}(x) I_{(0,\infty)}(\alpha) I_{(0,\infty)}(\lambda).$$

(a) Find the c.d.f. for X.
(b) Find the moment generating function for X.
(c) Use the moment generating function from part (b) to find $E[X]$ and $Var(X)$.
(d) Let $Y = e^{-(X-\alpha)}$. Find the c.d.f. for Y.

4.40. Let X be a continuous random variable defined on (a, b), where $-\infty < a < 0 < b < \infty$, with c.d.f. $F_X(x)$ and $E[X] < \infty$. Show that

$$\int_0^b [1 - F_X(t)] dt - \int_a^0 F_X(t) dt = E[X].$$

4.41. Let X be a continuous random variable with c.d.f.

$$\begin{aligned}
F_X(x) &= 0, \quad -\infty < x < 0, \\
&= \frac{x}{8}, \quad 0 \le x < 2, \\
&= \frac{x^2}{16}, \quad 2 \le x < 4, \\
&= 1, \quad x \ge 4.
\end{aligned}$$

(a) Find the p.d.f. for X.
(b) Obtain $E[X]$.

4.42. Let X be a continuous random variable with p.d.f.

$$f_X(x) = cx^2(1+x)I_{(-1,0)}(x)$$
$$= cx^2(1-x)I_{(0,1)}(x).$$

(a) Find the constant c that makes $f_X(x)$ a p.d.f.
(b) Find $E[X]$ and $Var(X)$.

4.43. Let X be a continuous random variable with p.d.f.

$$f_X(x) = \frac{1}{2\sigma}\phi(x/\sigma) + \frac{1}{2\sigma}\phi((x-1)/\sigma),$$

where $\phi(\cdot)$ is the p.d.f. for the standard normal distribution. Find the moment generating function for X.

4.44. Let $Y \sim NegBin\ (r, p)$ and $Z \sim Binom\ (n, p)$, where $0 < p < 1$ and r and n are positive integers. Show that

$$P(Y \le n) = P(Z \ge r).$$

4.45. Let $Y \sim Gamma\ (r, 1/\lambda)$ and $Z \sim Poisson\ (\lambda)$, where r is a positive integer and $\lambda > 0$. Show that

$$P(Y \le 1) = P(Z \ge r).$$

4.46. Let X be a continuous random variable with p.d.f.

$$f_X(x) = \theta(1+x)^{-(1+\theta)}I_{(0,\infty)}(x)I_{(2,\infty)}(\theta).$$

Find $E[X]$ and $Var(X)$.

4.47. Let X be a continuous random variable with c.d.f.

$$F_X(x) = e^{-e^{-x}}I_{(-\infty,\infty)}(x).$$

(This is called the *Type I Extreme Value Distribution*.)

(a) Find the p.d.f. for X.
(b) Obtain the moment generating function for X. Express your result in terms of the gamma function $\Gamma(\cdot)$.

4.48. Suppose that $\{\epsilon_t : t = 0, 1, \ldots\}$ is a sequence of independent and identically distributed random variables with mean 0 and variance σ^2. The sequence $\{X_t : t = 1, \ldots, n\}$ is generated according to the following model:

$$X_t = \epsilon_t - \theta\, \epsilon_{t-1},$$

where $-\infty < \theta < \infty$. Find $E[X_t]$ and $Var(X_t)$.

4.49. In some settings, the zero outcome for a Poisson random variable $Y \sim Poisson\,(\lambda)$ cannot be observed. This leads to a random variable X with the *0-truncated Poisson distribution* with p.d.f. given by

$$f_X(x) = P(X = x) = \frac{\lambda^x e^{-\lambda}}{x!P(Y > 0)} I_{\{1,2,\ldots\}}(x) I_{(0,\infty)}(\lambda).$$

(a) Obtain $P(Y > 0)$ to complete the formulation for $f_X(x)$.
(b) Find $E[X]$ and $Var(X)$.

4.50. Let X be a continuous random variable with p.d.f.

$$g_X(x) = p\, f_1(x) + (1 - p)f_2(x),$$

where $0 < p < 1$, $f_1(x)$ is a p.d.f. for a continuous distribution with mean μ_1, finite variance σ_1^2, and m.g.f. $M_1(t)$ and $f_2(x)$ is a p.d.f. for a continuous distribution with mean μ_2, finite variance σ_2^2, and m.g.f. $M_2(t)$. Find the moment generating function for X and use it to determine $E[X]$ and $Var(X)$.

Chapter 5
Joint Probability Distributions for Two Random Variables

In the previous two chapters, we discussed univariate random variables and properties of their probability distributions. Many statistical settings, however, involve more than a single variable. In this chapter, we introduce the concept of joint probability distributions for two or more random variables associated with the same experimental setting and discuss some of the important properties of such joint probability distributions.

5.1 Joint Probability Distributions of Two Variables

Consider an experiment for which the outcome is a pair of real numbers. Let X and Y represent these random outcomes, and let S represent the two-dimensional sample space or support for (X, Y).

Case 1 If the support S is finite or at most countably infinite, (X, Y) is a pair of **discrete random variables**.

Case 2 If the support S is not at most countably infinite, (X, Y) is a pair of **continuous random variables**.

5.1.1 Discrete Variables

Definition 5.1 The **joint probability function (p.d.f.) for a pair of discrete random variables (X,Y)** is a function $f_{X,Y}(x, y)$ that satisfies

(i) $f_{X,Y}(x, y) = 0 \quad \forall\, (x, y) \notin S$

(continued)

© Springer Nature Switzerland AG 2020
D. Wolfe, G. Schneider, *Primer for Data Analytics and Graduate Study in Statistics*,
https://doi.org/10.1007/978-3-030-47479-9_5

Definition 5.1 (continued)
(ii) For any $(x, y) \in S$, we have

$$P(X = x, Y = y) = f_{X,Y}(x, y) \geq 0 \quad \forall (x, y) \in S.$$

We call

$$P(X = x, Y = y), (x, y) \in S = f_{X,Y}(x, y) I_S(x, y)$$

the **joint probability distribution** for (X, Y), where $I_S(x, y)$ is the indicator function for the sample space S.

For discrete variables (X, Y) it follows from the additive property of probabilities over disjoint unions that

$$P\{(X, Y) \in A\} = \sum_{(x, y) \in A} f_{X,Y}(x, y) \quad \text{for any subset } A \subset S.$$

Example 5.1 Rolling a Pair of Fair Dice

Roll a pair of fair dice, and let $X =$ [the outcome on the first die] and $Y =$ [outcome on the second die]. Then

$$S = \{(x, y) : x = 1, \ldots, 6 \quad \text{and} \quad y = 1, \ldots, 6\}$$

and the joint probability function for (X, Y) is

$$f_{X,Y}(x, y) = \frac{1}{36} I_S(x, y).$$

Thus

$$P(X = Y) = \sum_{y=1}^{6} \sum_{x=y}^{y} \frac{1}{36} = \frac{6}{36} = \frac{1}{6}$$

and

$$P(X \text{ is even}, Y \text{ is odd}) = \sum_{y=1(2)}^{6} \sum_{x=2(2)}^{6} \frac{1}{36} = \frac{9}{36} = \frac{1}{4}.$$

Note that this last result also follows from the fact that the two events $\{X \text{ is even}\}$ and $\{Y \text{ is odd}\}$ are independent, each with probability ½, so that

$$P(X \text{ is even}, Y \text{ is odd}) = P(X \text{ is even}) P(Y \text{ is odd}) = \frac{1}{2} \frac{1}{2} = \frac{1}{4}.$$

Example 5.2 Dependent Discrete Variables

Let (X, Y) be discrete random variables with joint probability function:

$$f_{X,Y}(x,y) = \frac{x+y}{21} I_{\{x=1,2,3; y=1,2\}}(x,y).$$

Then

$$P(X = 2, Y = 1) = \frac{2+1}{21} = \frac{1}{7},$$

$$P(X = 3) = \sum_{y=1}^{2} \sum_{x=3}^{3} \frac{x+y}{21} = \frac{3+1}{21} + \frac{3+2}{21} = \frac{9}{21} = \frac{3}{7}$$

and

$$P(Y < X) = \sum_{x=2}^{3} \sum_{y=1}^{x-1} \frac{x+y}{21} = \frac{2+1}{21} + \frac{3+1}{21} + \frac{3+2}{21} = \frac{12}{21} = \frac{4}{7}.$$

As we shall see later, it is often useful to present a discrete joint probability function in simple tabular form when S contains only a small number of (x, y) pairs. Thus, the tabular form for the joint probability function in this example is given by the entries in the following table:

	$y = 1$	$y = 2$
$x = 1$	2/21	3/21
$x = 2$	3/21	4/21
$x = 3$	4/21	5/21

Example 5.3 Dealing a 13-Card Hand

Deal a 13-card hand from an ordinary 52-card deck of cards (no jokers). Let X and Y be the number of hearts and spades, respectively, in the hand. Then the joint probability function for (X, Y) is given by

$$f_{X,Y}(x,y) = \frac{\binom{13}{x}\binom{13}{y}\binom{26}{13-x-y}}{\binom{52}{13}}, \quad x = 0, 1, \ldots, 13; y = 0, 1, \ldots, 13;$$

such that $x + y \le 13$.

$$(5.1)$$

5.1.2   Continuous Variables

Case 2 S is not a countable set.

> **Definition 5.2** The **joint probability function (p.d.f.) for a pair of continuous random variables** (X, Y) is a function $f_{X,Y}(x, y)$ that satisfies
>
> (i) $f_{X,Y}(x, y) \geq 0 \ \forall \ (x, y) \in S$
> (ii) $f_{X,Y}(x, y) = 0, \ \forall \ (x, y) \notin S$
> (iii) For any $a < b$ and $c < d$, we have
>
> $$P(a < X < b, c < Y < d) = \int_c^d \int_a^b f_{X,Y}(x, y) dx dy.$$

Example 5.4 Joint Uniform Continuous Variables on the Unit Square
Let (X, Y) be continuous random variables with joint uniform p.d.f.:

$$f_{X,Y}(x, y) = 1 I_{\{0 < x < 1, 0 < y < 1\}}(x, y).$$

Then, we have

$$P\left(\frac{1}{2} < X < \frac{3}{4}, \frac{1}{2} < Y < 1\right) = \int_{1/2}^{1} \int_{1/2}^{3/4} dx\, dy = \int_{1/2}^{1} x\big|_{1/2}^{3/4} dy = \int_{1/2}^{1} \frac{1}{4} dy = \frac{1}{8}.$$

Think About It Could we compute the probability in Example 5.4 using only probabilities about the variables X and Y separately?

Example 5.5 Joint Uniform Continuous Random Variables on a Triangle
Let (X, Y) be continuous random variables with joint uniform p.d.f.:

$$f_{X,Y}(x, y) = 2 I_{\{0 < x < y < 1\}}(x, y).$$

Then, it follows that

$$P\left(\frac{1}{2} < X < \frac{3}{4}, \frac{1}{2} < Y < 1\right) = \int_{1/2}^{3/4} \int_{x}^{1} 2\, dy\, dx = \int_{1/2}^{3/4} 2y\big|_{x}^{1} dx$$

$$= \int_{1/2}^{3/4} 2(1 - x) dx = -(1 - x)^2 \big|_{1/2}^{3/4}$$

$$= -\frac{1}{16} + \frac{1}{4} = \frac{3}{16}.$$

Think About It Could we compute the probability in Example 5.5 using only probabilities about the variables X and Y separately? What is different about the settings for Examples 5.4 and 5.5?

Note There are situations in which one of a pair of random variables is discrete and the other is continuous, but we will not consider such settings in this text.

5.2   Marginal Probability Distributions

Definition 5.3 Let (X, Y) be random variables with joint p.d.f. $f_{X,Y}(x, y)$. The **marginal p.d.f.'s of X and Y** are given by

$$f_X(x) = \int_{-\infty}^{\infty} f_{X,Y}(x, y)\, dy \left[f_X(x) = \sum_y f_{X,Y}(x, y) \right]$$

and

$$f_Y(y) = \int_{-\infty}^{\infty} f_{X,Y}(x, y)\, dx \left[f_Y(y) = \sum_x f_{X,Y}(x, y) \right],$$

respectively.

Example 5.6 Dependent Discrete Variables
Let (X, Y) be discrete random variables with joint p.d.f.:

$$f_{X,Y}(x, y) = \frac{x + y}{21} I_{\{x=1,2,3;\, y=1,2\}}(x, y)$$

The marginal p.d.f. for X is then

$$f_X(x) = \sum_{y=1}^{2} \frac{x + y}{21} = \frac{2x + 3}{21} I_{\{1,2,3\}}(x)$$

and the marginal p.d.f. for Y is

$$f_Y(y) = \sum_{x=1}^{3} \frac{x + y}{21} = \frac{3y + 6}{21} I_{\{1,2\}}(y).$$

Another way to get these marginal p.d.f.'s is to put the joint p.d.f. in tabular form, as was done in Example 5.2, and sum across the columns to get $f_X(x)$ or down the rows to get $f_Y(y)$.

	y = 1	y = 2	$f_X(x)$
x = 1	2/21	3/21	5/21
x = 2	3/21	4/21	7/21
x = 3	4/21	5/21	9/21
$f_Y(y)$	9/21	12/21	

These sums appear in the "margins" of the table—hence, the name marginal probabilities.

Example 5.7 Dealing a 13-Card Hand

Deal a 13-card hand from the usual 52-card deck (no jokers), and let X and Y be the number of hearts and spades, respectively, in the hand. The joint p.d.f. for (X, Y) is given by Eq. 5.1 in Example 5.3. To obtain the marginal p.d.f.'s for X and Y, we COULD sum the joint p.d.f. over y and x, respectively. However, it is much easier in this case just to recognize that X and Y are simply the number of hearts and spades, respectively, in a 13-card hand from a 52-card deck with respective p.d.f.'s:

$$f_X(x) = \frac{\binom{13}{x}\binom{39}{13-x}}{\binom{52}{13}} I_{\{0,1,...,13\}}(x)$$

and

$$f_Y(y) = \frac{\binom{13}{y}\binom{39}{13-y}}{\binom{52}{13}} I_{\{0,1,...,13\}}(y).$$

Example 5.8 Joint Uniform Continuous Variables on the Unit Square

Let (X, Y) have joint p.d.f.:

$$f_{X,Y}(x,y) = 1 I_{(0,1)}(x) I_{(0,1)}(y).$$

The marginal p.d.f.'s are then

$$f_X(x) = \int\limits_0^1 1 dy = 1\!I_{(0,1)}(x) \quad \text{and} \quad f_Y(y) = \int\limits_0^1 1 dx = 1\!I_{(0,1)}(y).$$

Example 5.9 Joint Uniform Continuous Random Variables on a Triangle
Let (X, Y) have joint p.d.f.:

$$f_{X,Y}(x, y) = 2I_{\{0<x<y<1\}}(x, y)$$

The marginal p.d.f.'s are given by

$$f_X(x) = \int\limits_x^1 2 dy = 2(1-x)I_{(0,1)}(x) \quad \text{and} \quad f_Y(y) = \int\limits_0^y 2 dx = 2yI_{(0,1)}(y).$$

Think About It Now, do you see any differences between the joint uniform distribution on the unit square and the joint uniform distribution on a triangle? What do you think this means about obtaining joint probabilities from separate marginal probabilities?

Note In Example 5.6 we demonstrated how the marginal p.d.f.'s for discrete random variables can be obtained by summing across the rows or columns to obtain the marginal probabilities in the "margins." We can think of marginal p.d.f.'s for continuous random variables in a similar way, where integration over y (or x) plays the role of "collecting" all of the probability for a fixed x (y) value and depositing it on the x (y) axis (i.e., in the "margins").

5.3   Covariance and Correlation

Definition 5.4 Let (X, Y) have joint p.d.f. $f_{X,Y}(x, y)$, and let μ_X, μ_Y, σ_X^2, and σ_Y^2 be the respective means and variances. The **covariance between X and Y,** denoted by σ_{XY}, is defined by

$$\sigma_{XY} = E[(X - \mu_X)(Y - \mu_Y)] = E[XY] - \mu_X\mu_Y.$$

The **correlation between X and Y, denoted by** ρ_{XY}, is defined by

$$\rho_{XY} = \frac{\sigma_{XY}}{\sigma_X\sigma_Y}.$$

Example 5.10 Joint Uniform Continuous Variables on the Unit Square

Let (X, Y) have joint p.d.f.:

$$f_{X,Y}(x, y) = 1I_{(0,1)}(x)I_{(0,1)}(y).$$

Then

$$E[XY] = \int_0^1 \int_0^1 xy(1)\,dx\,dy = \int_0^1 \frac{x^2 y}{2}\Big|_{x=0}^{x=1} dy$$

$$= \int_0^1 \frac{y}{2}\,dy = \frac{y^2}{4}\Big|_{y=0}^{y=1} = \frac{1}{4}.$$

Using the marginal p.d.f's obtained in Example 5.8, we obtain

$$E[X] = E[Y] = \int_0^1 x\,dx = \frac{x^2}{2}\Big|_{x=0}^{x=1} = \frac{1}{2}$$

and

$$E[X^2] = E[Y^2] = \int_0^1 x^2\,dx = \frac{x^3}{3}\Big|_{x=0}^{x=1} = \frac{1}{3},$$

so that

$$\sigma_X^2 = \sigma_Y^2 = \frac{1}{3} - \left(\frac{1}{2}\right)^2 = \frac{1}{12}.$$

Thus,

$$\sigma_{XY} = \frac{1}{4} - \left(\frac{1}{2}\right)\left(\frac{1}{2}\right) = 0 \quad \text{and} \quad \rho_{XY} = \frac{0}{\sqrt{\frac{1}{12}}\sqrt{\frac{1}{12}}} = 0.$$

Think About It Suppose we had not previously obtained the marginal p.d.f.'s for X and Y in this example. Would we have to obtain these marginal distributions first in order to calculate $E[X]$, $E[Y]$, σ_X^2, and σ_Y^2?

Example 5.11 Joint Uniform Continuous Variables on a Triangle

Let (X, Y) have joint p.d.f.:

$$f_{X,Y}(x, y) = 2I_{\{0<x<y<1\}}(x, y).$$

Then

$$E[XY] = \int\limits_0^1 \int\limits_0^y 2xy\, dx\, dy = \int\limits_0^1 x^2 y \Big|_{x=0}^{x=y} dy = \int\limits_0^1 y^3 dy = \frac{y^4}{4}\Big|_{y=0}^{y=1} = \frac{1}{4}.$$

Using the marginal p.d.f.'s obtained in Example 5.9, we find

$$E[X] = \int\limits_0^1 2x(1-x)\, dx = \left[x^2 - \frac{2}{3}x^3\right]_{x=0}^{x=1} = 1 - \frac{2}{3} = \frac{1}{3},$$

$$E[X^2] = \int\limits_0^1 2x^2(1-x)\, dx = \left[\frac{2}{3}x^3 - \frac{2}{4}x^4\right]_{x=0}^{x=1} = \frac{2}{3} - \frac{2}{4} = \frac{1}{6},$$

$$E[Y] = \int\limits_0^1 2y^2 dy = \left[\frac{2}{3}y^3\right]_{y=0}^{y=1} = \frac{2}{3},$$

and

$$E[Y^2] = \int\limits_0^1 2y^3 dy = \left[\frac{1}{2}y^4\right]_{y=0}^{y=1} = \frac{1}{2}.$$

It follows that

$$\sigma_X^2 = \frac{1}{6} - \left(\frac{1}{3}\right)^2 = \frac{1}{18}, \quad \sigma_Y^2 = \frac{1}{2} - \left(\frac{2}{3}\right)^2 = \frac{1}{18}$$

and

$$\sigma_{XY} = \frac{1}{4} - \frac{1}{3}\left(\frac{2}{3}\right) = \frac{1}{36} > 0 \quad \Rightarrow \quad \rho_{XY} = \frac{\frac{1}{36}}{\sqrt{\frac{1}{18}}\sqrt{\frac{1}{18}}} = \frac{1}{2}.$$

Example 5.12 Dependent Discrete Variables
Let (X, Y) be discrete random variables with joint p.d.f.:

$$f_{X,Y}(x,y) = \frac{x+y}{21} I_{\{x=1,2,3;\, y=1,2\}}(x,y).$$

Then,

$$E[XY] = \sum_{x=1}^{3} \sum_{y=1}^{2} \frac{(x+y)}{21} xy$$

$$= \left[1\left(\frac{2}{21}\right) + 2\left(\frac{3}{21}\right) + 2\left(\frac{3}{21}\right) + 4\left(\frac{4}{21}\right) + 3\left(\frac{4}{21}\right) + 6\left(\frac{5}{21}\right) \right]$$

$$= \frac{24}{7} = 3.429.$$

Using the marginal distributions from Example 5.6, we find that

$$E[X] = 1\left(\frac{5}{21}\right) + 2\left(\frac{7}{21}\right) + 3\left(\frac{9}{21}\right) = \frac{46}{21} = 2.190,$$

$$E[Y] = 1\left(\frac{9}{21}\right) + 2\left(\frac{12}{21}\right) = \frac{11}{7} = 1.571$$

$$E[X^2] = 1^2\left(\frac{5}{21}\right) + 2^2\left(\frac{7}{21}\right) + 3^2\left(\frac{9}{21}\right) = \frac{38}{7},$$

and

$$E[Y^2] = 1^2\left(\frac{9}{21}\right) + 2^2\left(\frac{12}{21}\right) = \frac{19}{7}.$$

It follows that

$$\sigma_X^2 = \frac{38}{7} - \left(\frac{46}{21}\right)^2 = 5.429 - 4.728 = .631, \quad \sigma_Y^2 = \frac{19}{7} - \left(\frac{11}{7}\right)^2 = \frac{12}{49} = .245,$$

which leads to

$$\sigma_{XY} = \frac{24}{7} - \frac{46}{21}\left(\frac{11}{7}\right) = 3.429 - 3.442 = -.031 < 0$$

and

$$\rho_{XY} = \frac{-.013}{\sqrt{.631}\sqrt{.245}} = \frac{-.013}{.393} = -.033.$$

Note 1 Both the mean and variance depend on the measurement unit for the associated random variable, and the covariance between two random variables depends on the measurement units for both of the random variables. On the other hand, the correlation between two random variables does not depend on the measurement units for either variable; that is, the correlation is a unit-free quantity.

Note 2 The correlation between two random variables provides some information about the strength of any linear relationship between the variables. A positive value of ρ_{XY} corresponds to the tendency for larger values of X to be associated with larger values of Y, while a negative value of ρ_{XY} suggests that larger values of X tend to be associated with smaller values of Y. The numerical magnitude $|\rho_{XY}|$ is indicative of the strength of this positive or negative association, so that a value of ρ_{XY} close to zero corresponds to little or no linear relationship between the two variables. We should caution, however, that there could still be a non-linear relationship between two variables even if ρ_{XY} is 0, as you are asked to show in Exercise 5.17.

Mathematical Moment 7 Schwarz's Inequality
Let X and Y be random variables such that $E[XY] < \infty$, $E[X^2] < \infty$, and $E[Y^2] < \infty$. Then

$$E[XY] \le \sqrt{E[X^2]E[Y^2]},$$

where equality holds if and only if $Y = aX$ for some constant a.

Note 3 It follows immediately from Schwarz's inequality that $|\rho_{XY}| \le 1$, and $\rho_{XY} = 1$ is achieved only if there is a perfect linear relationship $Y = aX$ for some positive constant a while $\rho_{XY} = -1$ is achieved only if there is a perfect linear relationship $Y = aX$ for some negative constant a. As noted previously, the numerical value $|\rho_{XY}|$ provides an indication of the strength of any non-perfect linear relationship between X and Y.

5.4   Conditional Probability Distributions

In Chap. 2 we discussed the conditional probability $P(A|B)$ of an event A occurring given that a second event B has occurred. This concept extends quite naturally to conditional probabilities about a random variable Y given information about a second random variable X. These conditional probabilities depend entirely on the joint probability distribution for the pair (X, Y) and the marginal probability distribution for the conditioning random variable X.

Definition 5.5 Let (X, Y) have joint p.d.f. $f_{X,Y}(x, y)$ and marginal p.d.f.'s $f_X(x)$ and $f_Y(y)$, respectively. The **conditional p.d.f.'s of X given $Y = y$ and Y given $X = x$** are given by

$$f_{X|Y=y}(x \mid y) = \frac{f_{X,Y}(x, y)}{f_Y(y)} \quad \text{and} \quad f_{Y|X=x}(y \mid x) = \frac{f_{X,Y}(x, y)}{f_X(x)},$$

respectively.

The relationship between the joint, conditional, and marginal probability distributions is given by

$$f_{X,Y}(x,y) = f_{X|Y=y}(x|y)f_Y(y) = f_{Y|X=x}(y|x)f_X(x).$$

In the case of discrete distributions, this corresponds to the following easily recognizable relationship:

$$P(X = x, Y = y) = P(X = x|Y = y)P(Y = y) = P(Y = y|X = x)P(X = x).$$

If we define $A = \{X = x\}$, $B = \{Y = y\}$, this is simply what we learned in Chap. 2, namely,

$$P(A \cap B) = P(A|B)P(B) = P(B|A)P(A).$$

In Chap. 2 we also learned that two events A and B are independent if $P(A \cap B) = P(A)P(B)$. This concept also extends immediately to a pair of random variables (X, Y).

Definition 5.6 Let (X, Y) have joint p.d.f. $f_{X,Y}(x, y)$ and marginal p.d.f.'s $f_X(x)$ and $f_Y(y)$, respectively. The **two random variables X and Y are independent** if and only if

$$f_{X,Y}(x,y) = f_X(x)f_Y(y) \quad \text{for all } (x,y) \in S.$$

Random variables that are not independent are said to be **dependent.**

An immediate consequence of this relationship is that X and Y are independent if and only if

$$f_{X|Y=y}(x|y) = f_X(x) \quad \text{for all } (x,y),$$

so that $f_{X|Y=y}(x|y)$ does not depend on the observed value of y. Similarly, X and Y are independent if and only if

$$f_{Y|X=x}(y|x) = f_Y(y) \quad \text{for all } (x,y),$$

so that $f_{Y|X=x}(y|x)$ does not depend on the observed value of x.

Think About It What do these statements correspond to when X and Y are discrete random variables? Does this look familiar in the context of what we learned about independent events in Chap. 2?

Example 5.13 Dependent Discrete Variables

Let (X, Y) be discrete random variables with joint p.d.f.:

$$f_{X,Y}(x,y) = \frac{x+y}{21}I_{\{x=1,2,3;\,y=1,2\}}(x,y).$$

In Example 5.6, we saw that the marginal p.d.f.'s for X and Y are given by

$$f_X(x) = \sum_{y=1}^{2} \frac{x+y}{21} = \frac{2x+3}{21}\,I_{\{1,2,3\}}(x)$$

and

$$f_Y(y) = \sum_{x=1}^{3} \frac{x+y}{21} = \frac{3y+6}{21}\,I_{\{1,2\}}(y),$$

respectively.

It follows that the associated conditional p.d.f.'s are

$$f_{X|Y=y}(x|y) = \frac{\frac{x+y}{21}}{\frac{6+3y}{21}} = \frac{x+y}{6+3y}I_{\{1,2,3\}}(x), \quad y \in \{1,2\}, \text{fixed},$$

and

$$f_{Y|X=x}(y|x) = \frac{\frac{x+y}{21}}{\frac{3+2x}{21}} = \frac{x+y}{3+2x}I_{\{1,2\}}(y), \quad x \in \{1,2,3\}, \text{fixed}.$$

Since these conditional p.d.f.'s depend on the values of the conditioning variables, the two variables X and Y are not independent.

Example 5.14 Dependent Continuous Variables

Let (X, Y) be continuous random variables with joint p.d.f.:

$$f_{X,Y}(x,y) = e^{-y}I_{\{0<x<y<\infty\}}(x,y).$$

Then, the marginal p.d.f's are given by

$$f_X(x) = \int_x^\infty e^{-y}\,dy = -e^{-y}\Big|_x^\infty = e^{-x}I_{(0,\infty)}(x)$$

and

$$f_Y(y) = \int_0^y e^{-y}dx = xe^{-y}\Big|_0^y = ye^{-y}I_{(0,\infty)}(y).$$

(Note that the marginal distribution for X is *Gamma* $(\alpha = 1, \beta = 1)$ and the marginal distribution for Y is *Gamma* $(\alpha = 2, \beta = 1)$.)

The corresponding conditional p.d.f.'s are then

$$f_{X|Y=y}(x|y) = \frac{e^{-y}}{ye^{-y}} = \frac{1}{y}I_{(0,y)}, \quad y \in (0,\infty) \text{ fixed}$$

and

$$f_{Y|X=x}(y|x) = \frac{e^{-y}}{e^{-x}} = e^{-(y-x)}I_{(x,\infty)}(y), \quad x \in (0,\infty) \text{ fixed.}$$

Note Conditional probability distributions have all of the usual properties of probability distributions, such as means, variances, m.g.f.'s, etc.

Example 5.15 Dependent Discrete Variables

Let (X, Y) be discrete random variables with joint p.d.f.:

$$f_{X,Y}(x,y) = \frac{x+y}{21}I_{\{x=1,2,3;\,y=1,2\}}(x,y).$$

In Example 5.13, we saw that, for y fixed in $\{1,2\}$,

$$f_{X|Y=y}(x|y) = \frac{x+y}{6+3y}I_{\{1,2,3\}}(x).$$

The expected value of X given $Y = y$ fixed in $\{1,2\}$ is then given by

$$E[X|Y=y] = \sum_{x=1}^{3}\frac{x(x+y)}{6+3y} = \frac{1(1+y)+2(2+y)+3(3+y)}{6+3y} = \frac{14+6y}{6+3y}.$$

Thus, in particular,

$$E[X|Y=1] = \frac{14+6(1)}{6+3(1)} = \frac{20}{9} \quad \text{and} \quad E[X|Y=2] = \frac{14+6(2)}{6+3(2)} = \frac{13}{6}.$$

Combining these results with those from Example 5.12, we note that

$$E[X|Y=1] = \frac{20}{9} = 2.222 > E[X] = 2.190$$

and

$$E[X|Y=2] = \frac{13}{6} = 2.167 < E[X] = 2.190.$$

Think About It Do these comparisons suggest anything about the correlation between the two variables X and Y?

Example 5.16 Dependent Continuous Variables

In Example 5.14, we had, for y fixed in $(0, \infty)$,

$$f_{X|Y=y}(x|y) = \frac{1}{y} I_{(0,y)}(x).$$

It follows that

$$E[X|Y = y] = \int_0^y \frac{x}{y} dx = \frac{x^2}{2y} \Big|_{x=0}^{x=y} = \frac{y^2}{2y} - 0 = \frac{y}{2}, \quad y \text{ fixed in } (0, \infty)$$

and

$$E[X^2|Y = y] = \int_0^y \frac{x^2}{y} dx = \frac{x^3}{3y} \Big|_{x=0}^{x=y} = \frac{y^3}{3y} - 0 = \frac{y^2}{3}, \quad y \text{ fixed in } (0, \infty),$$

so that

$$Var(X|Y = y) = \frac{y^2}{3} - \left(\frac{y}{2}\right)^2 = \frac{y^2}{12}, \quad y \text{ fixed in } (0, \infty).$$

Example 5.17 Dependent Continuous Variables

In Example 5.14, we had, for fixed x in $(0, \infty)$,

$$f_{Y|X=x}(y|x) = e^{-(y-x)} I_{(x,\infty)}(y).$$

Using integration by parts, it follows that

$$E[Y|X = x] = \int_x^\infty y e^{-(y-x)} dy = e^x[-e^{-y} - ye^{-y}] \Big|_{y=x}^{y=\infty} = e^x[e^{-x} + xe^{-x}]$$

$$= 1 + x, \quad \text{for } x \text{ fixed in } (0, \infty).$$

Sometimes it is very natural to obtain conditional expectations and then use these conditional expectations to obtain corresponding marginal expectations without having to derive the complete marginal distributions. This very useful result is expressed in the following important property of iterated expectations, which we state without proof.

Result: Iterated Expectations

Let (X, Y) be a pair of random variables. Then it can be shown that

$$E[X] = E_Y\left[E_{X|Y}[X|Y]\right] \quad E[Y] = E_X\left[E_{Y|X}[Y|X]\right]$$

$$Var(Y) = E_X\left[Var_{Y|X}(Y|X)\right] + Var_X\left(E_{Y|X}[Y|X]\right)$$

and

$$Var(X) = E_Y\left[Var_{X|Y}(X|Y)\right] + Var_Y\left(E_{X|Y}[X|Y]\right).$$

Example 5.18 Dependent Continuous Variables
Let (X, Y) be continuous random variables with joint p.d.f.:

$$f_{X,Y}(x, y) = e^{-y}I_{\{0<x<y<\infty\}}(x, y).$$

In Example 5.14, we saw that the marginal distributions for X and Y are *Gamma* $(\alpha = 1, \beta = 1)$ and *Gamma* $(\alpha = 2, \beta = 1)$, respectively. From Examples 5.16 and 5.17, we also know that

$$E[X|Y = y] = \frac{y}{2} \quad \text{and} \quad E[Y|X = x] = 1 + x.$$

Using iterated expectations, it follows that

$$
\begin{aligned}
E[X] &= E_Y[E[X|Y = y]] \\
&= E_Y\left[\frac{Y}{2}\right] = \frac{1}{2}(2)(1) = 1, \quad \text{since } Y \sim Gamma(\alpha = 2, \beta = 1).
\end{aligned}
$$

This also follows immediately from the fact that $X \sim Gamma$ $(\alpha = 1, \beta = 1)$.
Similarly,

$$
\begin{aligned}
E[Y] &= E_X[E[Y|X = x]] = E_X[1 + X] = 1 + (1)(1) = 2, \\
&\text{since } X \sim Gamma(\alpha = 1, \beta = 1),
\end{aligned}
$$

once again agreeing with the more direct result from the fact that $Y \sim Gamma$ $(\alpha = 2, \beta = 1)$.

Also, using iterated expectations to obtain the variance of Y, we see that

$$Var(Y) = E_X[Var(Y|X = x)] + Var_X(E[Y|X = x]).$$

We previously found that $E[Y|X = x] = 1 + x$ and you are asked to show in Exercise 5.5 that $Var(Y|X = x) = 1$ for all x fixed in $(0, \infty)$. Thus, it follows that

$$Var(Y) = E_X[1] + Var_X(1 + X) = 1 + Var_X(X) = 1 + 1(1)^2 = 2,$$

since $X \sim Gamma\ (\alpha = 1, \beta = 1)$. This result also follows directly from the fact that $Y \sim Gamma\ (\alpha = 2, \beta = 1)$.

Now, you are no doubt wondering why we used iterated expectations in Example 5.18 when it was far easier to simply find the means and variances using the marginal distributions directly. We did this to illustrate the result in a setting where it could be verified independently from other results—particularly since we did not include a proof of the iterated expectation result. Our next example will illustrate that iterated expectations can be used effectively in more complicated settings where the marginal distributions are not so easily obtained.

Example 5.19 Roll a Fair Die and Flip a Fair Coin

Roll a fair six-sided die and let X be the outcome. Given the observed value of $X = x$, flip a fair coin x times, and let Y be the number of heads obtained. Then, from previous work, we know that X is uniformly distributed over the set $\{1, 2, 3, 4, 5, 6\}$ and that

$$E[X] = \frac{7}{2} \quad \text{and} \quad Var(X) = \frac{35}{12}.$$

Moreover, the conditional distribution of Y given $X = x$ is clearly $Binom\ (n = x, p = .5)$, with conditional mean and variance given by

$$E[Y|X = x] = \frac{x}{2} \quad \text{and} \quad Var(Y|X = x) = \frac{x}{4}.$$

Using iterated expectations, it follows that

$$E[Y] = E_X[E[Y|X = x]] = E_X\left[\frac{X}{2}\right] = \frac{1}{2}\left(\frac{7}{2}\right) = \frac{7}{4}$$

and

$$Var(Y) = E_X[Var(Y|X = x)] + Var_X(E[Y|X = x])$$

$$= E_X\left[\frac{X}{4}\right] + Var_X\left(\frac{X}{2}\right) = \left(\frac{7}{2}\right)\left(\frac{1}{4}\right) + \left(\frac{1}{4}\right)\left(\frac{35}{12}\right) = \frac{77}{48}.$$

These results could also be obtained by first finding the marginal distribution of Y. However, this is not nearly as straightforward as you might think, as you are asked to show in Exercise 5.1.

In our previous discussion about whether or not two random variables are independent or dependent, we have considered two approaches:

1. See if the joint p.d.f. can be obtained by multiplying the two marginal p.d.f.'s.
2. Check whether the conditional p.d.f. depends on the value of the conditioning variable.

Note that to use either of these approaches, we must first obtain at least one of the marginal p.d.f.'s before we can ascertain whether two variables are independent. In most situations, however, we do not need to actually obtain either of the marginal p.d.f.'s in order to decide if two random variables X and Y are independent, as the following result indicates.

Result: Independence and Factorization of Joint P.D.F.

Let (X, Y) be random variables with joint p.d.f. $f_{X,Y}(x, y)$. Then X and Y are independent if and only if we can write $f_{X,Y}(x, y)$ as a product of a nonnegative function of x alone times a nonnegative function of y alone AND the joint sample space for (X, Y) is a product space.

Example 5.20 Dependent Continuous Variables

Let (X, Y) be continuous random variables with joint p.d.f.:

$$f_{X,Y}(x, y) = 2I_{\{0<x<y<1\}}(x, y).$$

It follows from the previous result that X and Y are not independent variables since the support space $\{0 < x < y < 1\}$ is not a product space.

Example 5.21 Independent Continuous Variables

Let (X, Y) be continuous random variables with joint p.d.f.:

$$f_{X,Y}(x, y) = e^{-x-y}I_{(0,\infty)}(x)I_{(0,\infty)}(y).$$

Since the functional portion of the joint p.d.f. factors into the two parts e^{-x} times e^{-y} and the support space is a product space $(0, \infty) \times (0, \infty)$, it follows from the previous result that X and Y are independent variables.

Example 5.22 Dependent Continuous Variables

Let (X, Y) be continuous random variables with joint p.d.f.:

$$f_{X,Y}(x, y) = 2y^2 e^{-xy}I_{(0,\infty)}(x)I_{(0,1)}(y).$$

In this case, the sample space is indeed a product space $(0, \infty) \times (0, 1)$, but the functional portion of the joint p.d.f. cannot be factored into a function of x alone times a function of y alone, since the term e^{-xy} cannot be so factored.

Example 5.23 Dependent Discrete Variables

Let (X, Y) be discrete random variables with joint p.d.f.:

$$f_{X,Y}(x, y) = \frac{x+y}{21}I_{\{(1,1),(1,2),(2,1),(2,2),(3,1),(3,2)\}}(x, y).$$

In Example 5.13 we established that X and Y are dependent variables by showing that the conditional p.d.f.'s depend on the values of the conditioning variables. This involved first finding the marginal and then the conditional p.d.f.'s. However, with the factorization result, we can immediately conclude that X and Y are dependent variables since the functional form $(x + y)/21$ cannot be factored into a function of x alone times a function of y alone.

Example 5.24 Dealing a 13-Card Hand

Deal a hand of 13 cards from an ordinary deck of 52 playing cards (without the jokers) and let $X = $ (the number of hearts) and $Y = $ (the number of spades) in the hand. Then the joint distribution of (X, Y) is multivariate hypergeometric with joint p.d.f.:

$$f_{X,Y}(x,y) = \frac{\binom{13}{x}\binom{13}{y}\binom{26}{13-x-y}}{\binom{52}{13}} I_{\{0,1,2,...,13\}}(x) I_{\{0,1,2,...,13\}}(y) I_{\{0 \le x+y \le 13\}}(x,y).$$

Since the joint p.d.f. cannot be factored into a function of x alone times a function of y alone AND the joint sample space is not a product space, it follows from the factorization result that X and Y are dependent random variables. This is not a total surprise as we know that the number of spades we have in a hand must affect how many hearts we can have in the hand.

Example 5.25 Discrete Random Variables

Let (X, Y) be discrete variables with joint p.d.f. given by

$$f_{X,Y}(2,0) = f_{X,Y}(3,1) = \frac{\theta}{2} \quad \text{and} \quad f_{X,Y}(2,1) = f_{X,Y}(3,0) = \frac{1-\theta}{2}$$

for some value $\theta \in [0, 1]$.

For what θ values are X and Y independent? For X and Y to be independent, we must have

$$P(X = 2, Y = 0) = P(X = 2)P(Y = 0),$$

which requires that $\frac{\theta}{2} = \left(\frac{1}{2}\right)\left(\frac{1}{2}\right) = \frac{1}{4} \Leftrightarrow \theta = \frac{1}{2}$.

Moreover, when $\theta = \frac{1}{2}$, the joint p.d.f. simplifies to

$$f_{X,Y}(x,y) = \frac{1}{4} I_{\{2,3\}}(x) I_{\{0,1\}}(y),$$

which can be easily shown to factor into the product of the marginal p.d.f.'s for X and Y (and the joint sample space is a product space). Hence X and Y are independent variables if and only if $\theta = \frac{1}{2}$.

Result: Implications of Independence

Let X and Y be independent random variables. Let $a < b$ and $c < d$ be arbitrary constants, and let $g(X)$ and $h(Y)$ be arbitrary functions of X and Y alone, respectively. Then, the following facts follow immediately from the fact that the joint p.d.f. factors into the product of the marginal p.d.f.'s for independent random variables:

(i) $P(a < X < b, c < Y < d) = P(a < X < b) P(c < Y < d)$.

(ii) $E[g(X) h(Y)] = E[g(X)] E[h(Y)]$.

These expressions do not hold if X and Y are not independent.

Theorem 5.1 Independence and Correlation

Let X and Y be independent random variables. Then, if it exists, $Cov(X, Y) = 0$ and, hence, X and Y are uncorrelated.

Proof From Fact (ii) above, it follows that

$$Cov(X, Y) = E[(X - \mu_X)(Y - \mu_Y)] = E[(X - \mu_X)]E[(Y - \mu_Y)] = 0$$

$$\Rightarrow \ \rho_{XY} = \frac{Cov(X, Y)}{\sigma_X \sigma_Y} = 0, \text{ as well. } \blacksquare$$

However, the converse is not true, as the following example demonstrates.

Example 5.26 Dependent Random Variables with Zero Correlation

Let (X, Y) be discrete random variables with joint p.d.f.:

$$f_{X,Y}(x, y) = \frac{1}{3} I_{\{(0,0),(1,1),(2,0)\}}(x, y)$$

Clearly, X and Y are not independent, since the sample space is not a product space. However, we have

$$E[X] = 0\left(\frac{1}{3}\right) + 1\left(\frac{1}{3}\right) + 2\left(\frac{1}{3}\right) = 1 \quad E[Y] = 0\left(\frac{1}{3}\right) + 1\left(\frac{1}{3}\right) + 0\left(\frac{1}{3}\right) = \frac{1}{3}$$

and

$$E[XY] = 0(0)\left(\frac{1}{3}\right) + 1(1)\left(\frac{1}{3}\right) + 2(0)\left(\frac{1}{3}\right) = \frac{1}{3},$$

which implies that

$$Cov(X, Y) = E[XY] - E[X]E[Y] = \frac{1}{3} - 1\left(\frac{1}{3}\right) = 0.$$

Extensions to n Variables

Let the n random variables $X_1, \ldots, X_n$ have joint p.d.f. $f(X_1, \ldots, X_n)$ and marginal p.d.f.'s $f_1(x_1), \ldots, f_n(x_n)$, respectively. Then the random variables $X_1, \ldots, X_n$ are *mutually independent* if and only if

$$f(x_1, \ldots, x_n) = \prod_{i=1}^{n} f_i(x_i).$$

Implications of Mutual Independence

Let $X_1, \ldots, X_n$ be mutually independent random variables. The following two important results are immediate consequences of this mutual independence.

1. Let $a_i < b_i$, $i = 1, \ldots, n$, be constants. Then

$$P(a_1 < X_1 < b_1, a_2 < X_2 < b_2, \ldots, a_n < X_n < b_n) = \prod_{i=1}^{n} P(a_i < X_i < b_i).$$

2. Let $u_i(X_i)$ be a function of X_i alone, for $i = 1, \ldots, n$. Then

$$E\left[\prod_{i=1}^{n} u_i(X_i)\right] = \prod_{i=1}^{n} E[u_i(X_i)].$$

5.5 Exercises

5.1. Consider the joint distribution of (X, Y) described in Example 5.6. Find the marginal distribution of X, and use it directly to find $E[X]$ and $Var(X)$. How do you like the property of iterated expectations now?

5.2. Consider the setting of Example 5.13. Find an expression for the conditional expected value of Y given $X = x$, for x fixed in $\{1,2,3\}$.

5.3. In Example 5.14, note that the conditional probability distribution of X given $Y = y$ is nothing more than a uniform distribution over the interval $(0, y)$ for y fixed in $(0, \infty)$. Use this fact to obtain directly the conditional mean and variance for X given $Y = y$ in Example 5.16.

5.4. In Example 5.14, show that the conditional variance for Y given $X = x$ is equal to 1 for any x fixed in $(0, \infty)$.

5.5. Consider once again the conditional probability distribution of Y given $X = x$ for x fixed in $(0, \infty)$ obtained in Example 5.14. Find and identify the conditional probability distribution for the variable $Z = Y - x$, for x fixed in $(0, \infty)$. Use this fact to directly obtain the conditional mean and variance for Y given $X = x$, for x fixed in $(0, \infty)$.

5.6. Consider once again the setting for Example 5.18. Use iterated expectations to show that $Var(X) = 1$, in agreement with the fact that $X \sim Gamma$ $(\alpha = 1, \beta = 1)$.

5.7. Consider the joint distribution of (X, Y) described in Example 5.19. Find the marginal distribution of X, and use it directly to find $E[X]$ and $Var(X)$. How do you like the property of iterated expectations now?

5.8. Let (X, Y) be continuous random variables with joint p.d.f. given by

$$f_{X,Y}(x, y) = 1I_{(0,1)}(x)I_{(0,1)}(y).$$

(a) Find the marginal p.d.f.'s for X and Y.
(b) Are X and Y independent random variables? Justify your answer.

5.9. Let (X, Y) be continuous random variables with joint p.d.f given by

$$f_X(x, y) = e^{-y}I_{\{0<x<y<\infty\}}(x, y).$$

Are X and Y independent variables? Justify your answer.

5.10. Let (X, Y) be as defined in Example 5.24.

(a) Without trying to sum out the joint distribution of (X, Y), what are the marginal distributions of X and Y?
(b) Find an outcome (x, y) for which $f_{X, Y}(x, y) \neq f_X(x)f_Y(y)$ to deduce that X and Y are dependent random variables.
(c) Find the conditional probability functions for X given $Y = y$ and Y given $X = x$. Use these conditional probability functions to argue that X and Y are not independent variables.

Think About It: Do these conditional probability functions have intuitive interpretations?

5.11. Suppose we conduct independent Bernoulli trials with constant probability of success, $0 < p < 1$, until we get our first success. Let X be the number of Bernoulli trials needed to obtain this first success. Then we know that $X \sim Geom$ (p). Given the observed value of $X = x$, suppose we conduct x additional independent Bernoulli trials with the same parameter p and let Y denote the number of successes in these additional x Bernoulli trials.

(a) What is the conditional probability function for Y, given $X = x$?
(b) Obtain unconditional values for $E[Y]$ and $Var(Y)$.

5.12. Roll a pair of fair dice, one scarlet and one gray. Let X and Y represent the outcomes of the scarlet and gray dice, respectively.

(a) Find $P(X < 2Y)$.
(b) What is the probability distribution for the random variable $Z = Y - X$?

5.13. Let (X, Y) be a pair of continuous random variables with joint p.d.f.

$$f_{X,Y}(x, y) = 2y^2 e^{-xy} I_{(0,\infty)}(x) I_{(0,1)}(y).$$

(a) Are X and Y independent random variables? Justify your answer.
(b) Find the form of the conditional p.d.f. for X, given $Y = y$.
(c) Find $E[X]$.

5.14. A bowl contains the numbers 1 through 10 on separate folded pieces of paper. We draw one number at random from the bowl and let X denote the number drawn. Given $X = x$, we roll a fair die x times and let Y denote the number of these rolls for which the die exceeds 4.

(a) What is the probability distribution of X?
(b) Find an expression for the joint probability function for the pair (X, Y).
(c) Are X and Y independent random variables? Justify your answer.
(d) Find $E[Y]$ and $Var(Y)$.
(e) What is the probability that we drew the number 9 from the bowl if we observe a value of $Y = 8$?
(f) Find $P(X > Y)$.

5.15. Let X and Y be continuous random variables such that the conditional distribution of X given $Y = y$ is $n(y, 9)$ and the marginal distribution of Y is $n(5, 16)$.

(a) Are X and Y independent variables? Justify your answer.
(b) Obtain the joint p.d.f. for the pair (X, Y).
(c) Find $E[X]$ and $Var(X)$.
(d) Find the moment generating function for the marginal probability distribution of X, and use it to identify this marginal distribution. [Hint: Use iterated expectations.]

5.16. We have a coin for which $P(\text{head}) = p$, $0 < p < 1$. Consider the following three-step experiment:

   Step One: Flip the coin one time, and let X denote the number of heads obtained in this step.
   Step Two: Flip the same coin another $(X + 1)$ times, and let Y denote the number of heads obtained in this step.
   Step Three: Flip the same coin another $(X + Y + 1)$ times, and let Z denote the number of heads obtained in this step.

(a) Let T be the total number of heads obtained in the entire (all three steps) experiment. What is the sample space for T?
(b) Find the probability that you have observed at least one head after completion of all three steps of the experiment.
(c) Let T be as defined in part (a). Find $P(T = 3)$.
(d) Find $P(X = 0 | Z = 0)$.
(e) Let T be as defined in part (a). Find $E[T]$.

5.17. In Example 5.26, we described a setting where dependent random variables X and Y can still have zero correlation as measured by ρ_{XY}. In fact, ρ_{XY} is really only a measure of the *linear* relationship between X and Y. Construct an example where $\rho_{XY} = 0$ but where the two variables X and Y are perfectly dependent in the sense that if you know the value of X, you also know exactly the value of Y.

5.18. Let (X, Y) be continuous random variables with joint p.d.f.:

$$f_{X,Y}(x, y) = (x + y)I_{(0,1)}(x)I_{(0,1)}(y).$$

(a) Are X and Y independent variables? Justify your answer.
(b) Find the marginal p.d.f. for X.
(c) Find the conditional p.d.f. for Y given $X = x$.

5.19. Two players, A and B, independently flip fair coins until they each get their first head. Let X be the number of flips required by Player A, and let Y be the number of flips required by Player B.

(a) What are the probability distributions for X and Y?
(b) Obtain a numerical value for the probability that it takes Player B twice as many flips as it takes Player A.

5.20. Let (X, Y) be a pair of continuous random variables with joint p.d.f.:

$$f_{X,Y}(x, y) = \frac{1}{6\sqrt{2\pi}} y^{\frac{5}{2}} e^{-y} e^{-\frac{(x-y)^2}{2y}} I_{(-\infty,\infty)}(x) I_{(0,\infty)}(y).$$

(a) Are X and Y independent random variables? Justify your answer.
(b) Find the marginal p.d.f. for Y.
(c) Find the conditional p.d.f. for X given $Y = y$.
(d) Without first obtaining the marginal p.d.f. for X, find $E[X]$ and $Var(X)$.
(e) You could also use the alternative approach to first find the marginal p.d.f. of X and then use it directly to find $E[X]$ and $Var(X)$. Does this look like a reasonable approach compared to the route taken in part (d)?

5.21. Player A has probability p, $0 < p < 1$, of making a free throw, and Player B has probability q, $0 < q < 1$, of making a free throw. Each player shoots free throws "independently" until they make their first one.

(a) Find a closed form expression (not just a summation) for the probability that Player B needs at least as many shots as Player A to make her first free throw.
(b) If each player has the same probability of making a free throw (i.e., $p = q$), use the result in part (a) to find the probability that Player B needs exactly the same number of shots as Player A to make her first free throw.

5.22. Let (X, Y) be a pair of continuous random variables with joint p.d.f.:

$$f_{X,Y}(x,y) = \frac{1}{4!6^5 3!} y^3 e^{-\frac{x}{6}-\frac{y}{x}} I_{(0,\infty)}(x)\, I_{(0,\infty)}(y).$$

(a) Are X and Y independent variables? Justify your answer.
(b) Obtain the marginal p.d.f. for X.
(c) Obtain the conditional p.d.f. for Y given $X = x$.
(d) Find $E[Y]$.

5.23. Roll a fair six-sided die with sides numbered 1, 2, ..., 6, and let X be the number obtained. Given the value of $X = x$, roll a second fair die with x sides numbered 1, 2, ..., x, and let Y be the number obtained on the roll of this second die.

(a) Find the joint probability function for (X, Y).
(b) Are the random variables X and Y independent? Justify your answer.
(c) Obtain the values of $E[Y|X = x]$ and $Var(Y|X = x)$.
(d) Use the results in part (c) to find $E[Y]$ and $Var(Y)$.
(e) Find $Cov(X, Y)$. [Hint: Use appropriate iterated expectations to evaluate $E[XY]$.]

5.24. Suppose that the conditional distribution of X given $Y = y > 0$ is $n\left(\mu, \frac{1}{y}\right)$, with $-\infty < \mu < \infty$, and that the marginal distribution of Y is $Gamma$ (α, β), with $\alpha > 0$ and $\beta > 0$. Find $E[X]$ and $Var(X)$.

5.25. Let (X, Y) be random variables such that the conditional probability function of X given $Y = y$ is

$$f_{X|Y}(x|y) = \frac{y^x e^{-y}}{x!} I_{\{0,1,2,...\}}(x) I_{(0,\infty)}(y)$$

and

$$f_{X|Y}(0|0) = 1.$$

(a) Obtain expressions for $E[X|Y = y]$ and $Var(X|Y = y)$.
(b) Suppose the marginal distribution of Y has probability function:

$$f_Y(y) = \binom{n}{y} p^y (1-p)^{n-y} I_{\{0,1,\ldots,n\}} I_{(0,1)}(p).$$

Find $E[X]$ and $Var(X)$ as a function of p.

(c) Now suppose instead that the marginal distribution of Y has probability function

$$g_Y(y) = e^{-y} I_{(0,\infty)}(y).$$

Find the joint probability function for the pair (X, Y). Are X and Y independent variables? Justify your answer.

(d) Assuming the marginal distribution for Y as given in part (c), find and identify the marginal distribution of X.

5.26. Let (X, Y) be a pair of continuous random variables with joint p.d.f.:

$$f_{X,Y}(x,y) = 4xy \, I_{(0,1)}(x) I_{(0,1)}(y).$$

(a) Are X and Y independent variables? Justify your answer.
(b) Find the marginal p.d.f.'s for X and Y.

5.27. Let (X, Y) be a pair of continuous random variables with joint p.d.f.:

$$f_{X,Y}(x,y) = 8xy \, I_{\{0<x<y<1\}}(x,y).$$

(a) Are X and Y independent variables? Justify your answer.
(b) Find the marginal p.d.f.'s for X and Y.

5.28. Let X and Y be continuous random variables with conditional and marginal p.d.f.'s given by

$$f_{Y|X=x}(y|x) = \frac{3y^2}{x^3} \, I_{(0,x)}(y), \quad \text{for } x > 0 \text{ fixed}$$

and

$$f_X(x) = \frac{x^3 e^{-x}}{6} \, I_{(0,\infty)}(x),$$

respectively.

(a) Find the joint p.d.f. for (X, Y).
(b) Find the conditional p.d.f. of X given $Y = y > 0$.
(c) Find the moment generating function for the conditional distribution of X given $Y = y > 0$.

(d) Use the result in (c) to find $E[X^2|Y=y]$ for $y > 0$.

5.29. Holly is very popular among her classmates. Every year around the holiday season, she receives many gifts. Suppose the number of gifts Holly receives during the next holiday season, X, has a Poisson (λ) distribution. Holly also likes to re-give gifts that she receives, especially those that she thinks would fit others better. Let Y be the number of gifts that Holly re-gives during the next holiday season, and assume that, conditional on $X = x$, Y has a *Binom* (x, p) distribution.

(a) What is the joint probability function for (X, Y)?
(b) Are X and Y independent variables? Justify your answer.
(c) Derive the moment generating function for Y.
(d) Find $E[Y]$ and $Var(Y)$.

5.30. Let (X, Y) be discrete random variables with the joint probability function

$$f_{X,Y}(x,y) = \left[n!y^x\left(pe^{-1}\right)^y(1-p)^{n-y}\right]/[y!(n-y)!x!]I_{\{0,1,2,\dots\}}(x)I_{\{0,1,2,\dots,n\}}(y)$$

for $0 < p < 1$.

(a) Obtain the marginal probability function for Y and find $E[Y]$ and $Var(Y)$.
(b) Find the conditional probability function of X, given $Y = y$.
(c) Are X and Y independent variables? Justify your answer.
(d) From (b), obtain the values of $E[X|Y=y]$ and $Var(X|Y=y)$.
(e) Find $E[X]$ and $Var(X)$.

5.31. Let (X, Y) be discrete random variables with joint probability function as given in the following table:

(x,y)	$(0,0)$	$(0,1)$	$(1,0)$	$(1,1)$	$(2,0)$	$(2,1)$
$f_{X,Y}(x,y)$	$1/18$	$3/18$	$4/18$	$3/18$	$6/18$	$1/18$.

(a) Find the marginal probability functions for X and Y.
(b) Obtain the conditional probability functions for X, given $Y = y$, and for Y, given $X = x$.
(c) Find $E[X|Y=y]$ and $E[Y|X=x]$.

5.32. Let (X, Y) be continuous random variables such that the conditional p.d.f. for Y given $X = x$ and marginal p.d.f. for X are given by

$$f_{Y|X=x}(y|X=x) = e^{x-y}I_{\{0<x<y<\infty\}}(x,y)$$

and

$$f_X(x) = 2e^{-2x} I_{(0,\infty)}(x),$$

respectively.

(a) Are X and Y independent? Justify your answer.
(b) Find $E[Y|X = x]$.
(c) Find $E[Y]$ without first finding the marginal p.d.f. for Y.
(d) Find the marginal p.d.f. for Y.

5.33. Let (X, Y) be discrete random variables with joint p.d.f. given in the following table:

		x		
		1	2	3
	2	1/12	1/6	1/12
y	3	1/6	0	1/6
	4	0	1/3	0.

(a) Find the marginal c.d.f. for X and the marginal c.d.f. for Y.
(b) Find $E[X]$ and $E[Y]$.
(c) Find $Cov(X, Y)$. Are X and Y independent? Justify your answer.
(d) Find the conditional probability function for Y given $X = 2$.
(e) Find the conditional probability function for X given $Y = 4$.
(f) Find the probability function for the random variable $V = X - Y$.

5.34. Let X denote the number of heads that occur in three flips of a coin for which $p = P(\text{head})$ on each flip, $0 < p < 1$. Given $X = x \in \{0, 1, 2, 3\}$, consider a second experiment consisting of flipping the same coin until an additional $(x + 1)$ heads occur. Let Y denote the number of tails that occur **in this second experiment** before we observe the additional $(x + 1)$ heads.

(a) Find the form of the joint probability function for (X, Y).
(b) If the coin is fair and we observe $Y = 2$, what is the probability that we had obtained $X = 2$ heads in the three initial coin tosses?

5.35. Let X_1, X_2 be independent random variables, each with the same continuous probability distribution with p.d.f.:

$$f_X(x) = 2x \, I_{(0,1)}(x).$$

(a) Find $P(\max\{X_1, X_2\} \leq 0.5)$.
(b) Find $P(X_1 + X_2 \leq 1.5)$.

5.36. Let (X, Y) be continuous random variables with joint p.d.f.:

$$f_{X,Y}(x, y) = 2Ce^{-Cx} I_{\{0<y<x<\infty\}}(x, y),$$

where C is a constant.

(a) Determine the constant C so that $f_{X,Y}(x, y)$ is a proper joint p.d.f.
(b) Without any further calculations, state whether X and Y are independent variables. Justify your answer.
(c) Derive the marginal p.d.f. of Y.
(d) Obtain the conditional p.d.f. of X given $Y = y$.
(e) Find $E[X|Y = 1]$.

5.37. Suppose the time (in minutes) spent by a customer in a post office waiting to be served, X, and the total time spent in the post office, Y, have a joint p.d.f.:

$$f_{X,Y}(x, y) = 8(y - x)e^{-2y} I_{\{0<x<y<\infty\}}(x, y).$$

You may assume that the customer leaves the post office immediately after being served.

(a) Find the marginal p.d.f. for Y.
(b) Find the conditional p.d.f. of $X|Y$ and its expectation.
(c) Find the expected value of X without first finding its marginal p.d.f.
(d) Find the probability that the time that a customer is being served is longer than her waiting time.

5.38. Let X be a continuous random variable with p.d.f.:

$$f_X(x) = 2x(x^2 - 1)e^{-(x^2-1)} I_{(1,\infty)}(x).$$

Conditional on $X = x > 1$, the p.d.f. of the continuous random variable Y is

$$f_{Y|X=x}(y|x) = 2e^{-2(y-x^2+1)} I_{\{x^2-1,\infty)}(y).$$

(a) Are X and Y independent? Justify your answer.
(b) Find the conditional expected value of Y given $X = x$, for any given $x > 1$.
(c) Find the expected value of Y without first finding its marginal p.d.f.
(d) What is the probability that $X^2 - 1 \leq Y \leq 2X - 2$? [Hint: Draw a picture.]

5.39. For each $i = 1, 2, \ldots, 7$, let X_i be the number of meteorites of any type striking the earth on day i of this week, and let Y_i be the number of iron meteorites striking the earth on day i of this week. Assume that $X_1, \ldots, X_7$ are independent and identically distributed *Poisson* (20) variables and that the conditional distribution of Y_i given $X_i = x$ is *Binom* $(x, 0.8)$ for each $x = 0, 1, \dots$

(a) Give the joint probability function for (X_1, Y_1).
(b) Calculate $E[Y_1]$ and $Var(Y_1)$.
(c) Are X_1 and Y_1 independent? Justify your answer.
(d) Find the moment generating function for the total number of meteorites of any type striking the earth anytime this week.

5.40. Suppose that

$$Y|X \sim Poisson(g(\beta, X))$$

and

$$X \sim n(\mu, \sigma^2).$$

(a) Suppose $g(\beta, X) = e^{\beta X}$. What is $E[Y]$?
(b) Suppose $g(\beta, X) = 10^{\beta X}$. Find a lower bound for $E[Y]$ that is greater than 0. [Hint: Consider Jensen's inequality, which states that if Z is a random variable and $\varphi(\cdot)$ is a convex function, then $\varphi(E(Z)) < E[\varphi(Z)]$.]

5.41. Let $X_1, \dots, X_n$ be independent and identically distributed Bernoulli random variables with $P(X_1 = 1) = p$.

(a) Find $E\left[\prod_{i=1}^{n} (X_i + 1) \right]$.

(b) Find $E\left[X_1 \mid \sum_{i=1}^{n} X_i = k \right]$, where k is an integer between 0 and n, inclusive. [Hint: Remember that X_1 can only take on the two values 0 and 1.]

5.42. Suppose that $\{\epsilon_t : t = 0, 1, \dots\}$ is a sequence of independent and identically distributed random variables with mean 0 and variance σ^2. The sequence $\{X_t : t = 1, \dots, n\}$ is generated according to the following model:

$$X_t = \epsilon_t - \theta \, \epsilon_{t-1},$$

for some $\theta > 0$. Show that X_t and X_{t+k} are uncorrelated for any $k > 1$.

5.43. Let $X \sim Gamma (\propto, \beta)$, where $\beta > 0$ and $\propto$ is a **positive integer**. Suppose that the conditional distribution of Y given $X = x$ is *Poisson* (x). Use moment generating functions and iterated expectations to show that the unconditional distribution of X is $NegBin \left(\propto, \frac{1}{\beta+1} \right)$.

5.44. Consider a subject who walks into a clinic today, at time t, and is diagnosed with a certain infectious disease. At the same time t, a diagnostic measurement, Z_0, of the severity of the disease is obtained. Let S be the unknown date in the past when the subject initially became infected. We are interested in the time $Y_0 = t - S$ from initial infection until detection. Assume that the conditional distribution of Z_0 given $Y_0 = y_0$ is $n\left(\mu + \beta y_0, \sigma^2\right)$, where $\beta > 0$, and μ and σ^2 are the mean and variance, respectively, of Z_0 in the population of people without the disease. Here, βy_0 represents the mean increase in Z_0 for infected subjects over the time period y_0. It will be convenient to rescale the problem by introducing $Z = \frac{Z_0 - \mu}{\sigma}$ and $Y = \frac{\beta Y_0}{\sigma}$.

(a) Show that the conditional distribution of Z given $Y = y$ is $n(y, 1)$.
(b) Suppose that $Y \sim Exp(1/\lambda)$ for some $\lambda > 0$. Show that the conditional distribution of Y given $Z = z$ has p.d.f.

$$f_{Y|Z}(y|z) = (2\pi)^{-\frac{1}{2}} c^{-1} e\left\{-\frac{1}{2}[y-(z-\lambda)]^2\right\} I_{(0,\infty)}(y),$$

with

$$c = \Phi(z - \lambda),$$

where $\Phi(\cdot)$ is the c.d.f. for the standard normal distribution. [Note: This p.d.f is called the *truncated normal density*.]
(c) Find the conditional p.d.f. for Y_0 given $Z_0 = z_0$.
(d) Compute $E[Y|Z = z]$.

5.45. Let X and Y be random variables with c.d.f.'s $F_X(x)$ and $G_Y(y)$, respectively. Define the function:

$$H_{X,Y}(x, y) = F_X(x)G_Y(y)[1 + \alpha\{1 - F_X(x)\}\{1 - G_Y(y)\}]I_{[-1,1]}(\alpha).$$

(a) Show that $H_{X,Y}(x, y)$ is a valid bivariate c.d.f. for the pair (X, Y).
(b) Suppose that (X, Y) has joint c.d.f. $H_{X,Y}(x, y)$. Show that the marginal c.d.f for X is $F_X(x)$.
(c) Find the marginal c.d.f. for Y.
(d) Find the joint p.d.f. for the pair (X, Y).
(e) Assuming still that the (X, Y) pair is jointly distributed with c.d.f. $H_{XY}(x, y)$, further assume that, marginally $X \sim Exp(\lambda)$ and $Y \sim Exp(\lambda)$. Find $\sigma_{XY} = Cov(X, Y)$.
(f) Assuming the setup in part (d), find the mean and variance of the variable $W = \frac{1}{2}[X + Y]$.
(g) Assuming the setup in part (d), find a value of α for which X and Y are independent. Be sure to justify your answer.

5.46. Let x and y be real numbers, $x \vee y = max(x, y)$ and $x \wedge y = min(x, y)$. Let X and Y be random variables whose expectations exist.

(a) Prove that $x + y = (x \vee y) + (x \wedge y)$.
(b) Write an expression for $E[X \vee Y]$ in terms of $E[X]$, $E[Y]$, and $E[X \wedge Y]$.

(c) Prove that $|x - y| = (x \vee y) - (x \wedge y)$.

(d) Write an expression for $E[|X - Y|]$ in terms of $E[X]$, $E[Y]$, and $E[X \wedge Y]$.

(e) Let X and Y be independent random variables such that $X \sim Exp(\beta_X)$ and $Y \sim Exp(\beta_Y)$. Find $E[|X - Y|]$.

5.47. Let X and Y be continuous random variables such that the conditional p.d.f. for Y given $X = x$ is given by

$$f_{Y|X=x}(y|x) = \frac{cy}{x^2} I_{(0,x)}(y) \quad \text{for } x \text{ fixed in } (0, 1) \text{ and some constant } c.$$

(a) Find the constant c so that $f_{Y|X=x}(y|x)$ is a p.d.f.

(b) Are X and Y independent variables? Justify your answer.

(c) Find $E[Y|X = x]$ and $Var(Y|X = x)$.

Now suppose that the marginal p.d.f. for X is given by

$$f_X(x) = gx^4 I_{(0,1)}(x) \quad \text{for some constant } g.$$

(d) Find the constant g so that $f_X(x)$ is a p.d.f.

(e) Without obtaining the form of the marginal p.d.f. for Y, find $E[Y]$ and $Var(Y)$.

(f) Find the joint p.d.f. for the pair (X, Y), and obtain the marginal p.d.f. for Y from it.

(g) Find $Cov(X, Y)$.

(h) Compute $P(0.5 < X < .75, .6 < Y < .9)$.

5.48. Choose a point at random from the interval $(1, 10)$, and let the random variable X correspond to that point. Given $X = x$, generate a second random variable $Y \sim Exp\left(\frac{1}{x}\right)$.

(a) Are X and Y independent random variables? Justify your answer.

(b) Find $E[Y|X = x]$ and $Var(Y|X = x)$.

(c) Without obtaining the form of the marginal p.d.f. for Y, find $E[Y]$ and $Var(Y)$.

(d) Find the joint p.d.f. for the pair (X, Y).

Think About It: You could also obtain $E[Y]$ and $Var(Y)$ by first finding the marginal p.d.f. for Y and then using it directly to find these quantities. Want to try that? Check it out.

5.49. Consider two bowls filled with colored balls: Bowl 1 contains 7 red balls and 3 white balls; Bowl 2 contains 6 red balls and 4 white balls. We draw two balls from Bowl 1, record their colors, and place them in Bowl 2. Then, we draw two balls from the updated Bowl 2 and record their colors. Let X denote the number of red balls drawn from Bowl 1, and let Y denote the number of red balls drawn from Bowl 2.

(a) Are X and Y independent variables? Justify your answer.

(b) Find the joint probability distribution for the pair (X, Y).

(c) Find $Cov(X, Y)$.

(d) Find the marginal p.d.f.'s for X and Y.

(e) Find the conditional p.d.f. for Y given $X = x$ for each possible value of $x = 0$, 1, 2.

(f) Find $P(X = 2|Y = 2)$.

5.50. Let $X \sim \text{Gamma}(\alpha, \beta)$, with $\alpha > 0$ and $\beta > 0$. Suppose the conditional distribution of Y given $X = x$ is $n(x, x)$.

(a) Are X and Y independent random variables? Justify your answer.

(b) Without obtaining the form of the marginal p.d.f. for Y, find $E[Y]$ and $Var(Y)$.

(c) Find the joint p.d.f. for the pair (X, Y).

Think About It: You could also obtain $E[Y]$ and $Var(Y)$ by first finding the marginal p.d.f. for Y and then using it directly to find these quantities. Want to try that? Check it out.

5.51. Let (X, Y) be continuous random variables with joint p.d.f. given by

$$f_{(X,Y)}(x,y) = 6 \left[x^2 y + xy^2 \right] I_{\{0<x<y<1\}}(x,y).$$

(a) Are X and Y independent variables? Justify your answer.

(b) Find the marginal p.d.f.'s for X and Y.

(c) Find the conditional p.d.f. for X given $Y = y \in (0, 1)$. Calculate $E[X|Y = y]$ and $Var(X|Y = y)$.

(d) Find the conditional p.d.f. for Y given $X = x \in (0, 1)$. Calculate $E[Y|X = x]$ and $Var(Y|X = x)$.

(e) Find $Cov(X, Y)$.

Chapter 6
Probability Distribution of a Function of a Single Random Variable

In the previous five chapters, we have developed the concept of a random variable and presented some of the important properties of random variables and their probability distributions. Many times, in practice, however, it is not simply the measured random variable that is of interest to us. Rather, it is some function of that random variable that is of primary concern. In this chapter, we discuss three distinct techniques, each of which is valuable in different situations, for obtaining probability distributions for functions of a single random variable. We will expand on these techniques in Chap. 7 when we discuss the all-important topic of sampling distributions.

Let X be a random variable with p.d.f. $f_X(x)$, m.g.f. $M_X(t)$, and c.d.f. $F_X(x)$. Let $Y = u(X)$ be a function of X. We discuss three basic approaches for deriving the probability distribution of Y:

1. Change of variable technique
2. Moment generating function technique
3. Distribution function technique

6.1 Change of Variable Technique

We start with the change of variable technique, by far the most commonly used approach for finding the probability distribution of $Y = u(X)$. It is, in fact, the default approach that will always work, but for some settings it can be more difficult to use than one or both of the other two techniques.

In the change of variable approach, we directly find the p.d.f., say $g_Y(y)$, for Y. Assume for our purposes in this text that the function $u(\cdot)$ has a unique inverse $x = w(y)$. We now consider discrete and continuous variables separately.

© Springer Nature Switzerland AG 2020
D. Wolfe, G. Schneider, *Primer for Data Analytics and Graduate Study in Statistics*,
https://doi.org/10.1007/978-3-030-47479-9_6

Theorem 6.1 Change of Variable for a Discrete Variable

Let S be the support space for the discrete random variable X. Then, the p.d.f. for $Y = u(X)$ is given by

$$g_Y(y) = f_X(w(y))I_{\{u(S)\}}(y),$$

where $T = u(S)$ is the image of the X support space S under the transformation $u(\cdot)$.

Proof For $y \in T = u(S)$, we have

$$g_Y(y) = P(Y = y) = P(u(X) = y) = P(X = w(y)) = f_X(w(y)). \quad \blacksquare$$

Example 6.1 Binomial Distribution

Let $X \sim Binom\ (n, p)$ and set $Y = u(X) = n - X$. The unique inverse function is $x = w(y) = n - y$. The p.d.f. for X is given by

$$f_X(x) = \binom{n}{x}p^x(1 - p)^{n-x}I_{\{0,1,\dots,n\}}(x).$$

It follows from Theorem 6.1 that the p.d.f. for Y is given by

$$g_Y(y) = f_X(w(y)) = f_X(n - y) = \binom{n}{n - y}p^{n-y}(1 - p)^{n-(n-y)}I_{\{0,1,\dots,n\}}(n - y)$$

$$= \binom{n}{y}(1 - p)^y p^{n-y}I_{\{0,1,\dots,n\}}(y),$$

which implies that $Y \sim Binom\ (n, 1 - p)$. This is not surprising since Y is just the number of "failures" among n Bernoulli trials with constant probability of "success" p.

Example 6.2 Geometric Distribution

Let X be a $Geom\ (p = .5)$ variable with p.d.f.:

$$f_X(x) = \left(\frac{1}{2}\right)^x I_{\{1,2,3,\dots\}}(x).$$

Let $Y = u(X) = X^3$. The unique inverse function is $x = w(y) = \sqrt[3]{y}$ and it follows from Theorem 6.1 that the p.d.f. for Y is given by

$$g_Y(y) = f_X(w(y)) = f_X(\sqrt[3]{y}) = \left(\frac{1}{2}\right)^{\sqrt[3]{y}} I_T(y),$$

where $T = u(\{1, 2, 3, \dots\}) = \{1, 8, 27, \dots\}$ is the transformed sample space for $Y = X^3$.

Theorem 6.2 Change of Variable for a Continuous Variable
Let S be the support space for the continuous random variable X. Then, the p.d.f. for $Y = u(X)$ in this continuous setting is given by

$$g_Y(y) = f_X(w(y)) \mid w'(y) \mid I_{\{u(S)\}}(y),$$

where $w'(y) = \frac{dw(y)}{dy}$ and $T = u(S)$ is once again the image of the X support space S under the transformation $u(\cdot)$.

Proof While the result is true for any function $u(\cdot)$ with unique inverse $w(y)$, we provide the details of the proof only for the case of a monotone function. Let $a < b$ be such that $(a, b) \subset u(S)$. Then, if $u(\cdot)$ is a monotone function, it follows that

$$P(a < Y < b) = P(a < u(X) < b) = P(w(a) < X < w(b)) = \int_{w(a)}^{w(b)} f_X(x)dx.$$

Making the change of variable $y = u(x)$ in the integral, with the inverse function $x = w(y)$, we obtain

$$P(a < Y < b) = \int_a^b f_X(w(y)) \mid w'(y) \mid dy = \int_a^b g_Y(y)dy,$$

since $u(w(a)) = a$ and $u(w(b)) = b$, which yields the result. ∎

Example 6.3 Gamma Distribution
Let X be a continuous random variable with p.d.f.:

$$f_X(x) = \theta x^{\theta-1} I_{(0,1)}(x), \quad \text{with } 0 < \theta < 1.$$

(As we shall see later, this is a special case of a class of distributions known as the *Beta* distributions.) Set $Y = u(X) = -2\theta\ln(X)$. The unique inverse function is $x = w(y) = e^{-\frac{y}{2\theta}}$, and it follows that

$$w'(y) = -\frac{1}{2\theta}e^{-\frac{y}{2\theta}} \Rightarrow \mid w'(y) \mid = \frac{1}{2\theta}e^{-\frac{y}{2\theta}}$$

and the transformed sample space is

$$T = u(S) = \{0 < y < \infty\}.$$

Hence, from Theorem 6.2, the p.d.f. for Y is given by

$$g_Y(y) = f_X(w(y))|w'(y)| = \theta(e^{-\frac{y}{2\theta}})^{\theta-1} \left[\frac{1}{2\theta}e^{-\frac{y}{2\theta}}\right]$$

$$= \frac{1}{2}e^{-\frac{y}{2}}I_{(0,\infty)}(y).$$

$\Rightarrow Y \sim Gamma(\alpha = 1, \beta = 2)$, or $\chi^2(2)$, or exponential with $\lambda = 2$.

Example 6.4 Continuous Variable

Let X be a continuous random variable with p.d.f.:

$$f_X(x) = 4x^3\, I_{(0,1)}(x).$$

Set $Y = u(X) = X^4$. Then, the unique inverse function is $x = w(y) = \sqrt[4]{y}$, and it follows that

$$w'(y) = \frac{1}{4}y^{-\frac{3}{4}} \Rightarrow |w'(y)| = \frac{1}{4}y^{-\frac{3}{4}}$$

and the transformed sample space is

$$T = u(S) = \{0 < y < 1\}$$

Hence, from Theorem 6.2, the p.d.f. for Y is given by

$$g_Y(y) = f_X(w(y))|w'(y)| = 4\left(\sqrt[4]{y}\right)^3 \left[\frac{1}{4}y^{-\frac{3}{4}}\right] = 1\, I_{(0,1)}(y).$$

Thus, $Y = X^4 \sim Unif\,(0, 1)$. As we shall see later, this is a special case of a very important result called the Probability Integral Transformation that connects all continuous distributions to the $Unif\,(0, 1)$ distribution!

6.2   Moment Generating Function Technique

The moment generating function technique is not useful for every situation where we want the distribution of a function of a random variable, but when it is applicable, it provides an elegant way to establish such results.

The moment generating function for $Y = u(X)$ is given by $M_Y(t) = E_Y[e^{tY}]$. Looking at this expression, it appears that we would already have to know the probability distribution of Y in order to find $E_Y[e^{tY}]$, thereby defeating the purpose of this approach. However, we aren't actually looking directly for the entire distribution of Y, just the specific expectation $E_Y[e^{tY}]$. Moreover, we can find that expectation either by using the distribution of Y directly (which we do not know) or indirectly from the fact that it is also an expectation with respect to the known distribution of X, namely,

$$M_Y(t) = E_Y[e^{tY}] = E_x[e^{tu(X)}].$$

Thus, if we can obtain an expression for $M_Y(t) = E_X[e^{tu(X)}]$ *and* identify the probability distribution associated with $M_Y(t)$ (remember there is a unique one-to-one correspondence between a probability distribution and its moment generating function), we can specify the corresponding probability distribution for Y.

Example 6.5 Binomial Distribution
Let $X \sim Binom\,(n, p)$ and set $Y = n - X$. The moment generating function for Y is given by

$$M_Y(t) = E_Y[e^{tY}] = E_X[e^{t(n-X)}] = e^{tn}\,E[e^{-tX}]$$
$$= e^{tn}M_X(-t) = e^{tn}[pe^{-t} + (1-p)]^n$$
$$\Rightarrow M_Y(t) = [e^t(pe^{-t} + (1-p))]^n = [p + (1-p)e^t]^n,$$

which we recognize as the moment generating function for the $Binom\,(n, 1 - p)$ distribution. Hence, $Y \sim Binom\,(n, 1 - p)$, as we saw previously in Example 6.1.

Example 6.6 Normal Distribution
Let $X \sim n\,(\mu, \sigma^2)$ and set $Y = \frac{X - \mu}{\sigma}$. Then the moment generating function for Y is given by

$$M_Y(t) = E_Y[e^{tY}] = E_X\left[e^{t\left(\frac{X-\mu}{\sigma}\right)}\right] = e^{-\frac{t\mu}{\sigma}}E_X\left[e^{\frac{tX}{\sigma}}\right] = e^{-\frac{t\mu}{\sigma}}M_X\left(\frac{t}{\sigma}\right)$$
$$\Rightarrow M_Y(t) = e^{-\frac{t\mu}{\sigma}}e^{\frac{t\mu}{\sigma} + \frac{\sigma^2\left(\frac{t}{\sigma}\right)^2}{2}} = e^{\frac{t^2}{2}},$$

which we recognize as the moment generating function for the standard normal distribution. Hence, $Y \sim n\,(0, 1)$.

Example 6.7 Gamma Distribution
Let X be a continuous random variable with p.d.f.:

$$f_X(x) = \theta x^{\theta-1}I_{(0,1)}(x), \quad \text{with } 0 < \theta < \infty.$$

Let $Y = -2\theta\ln(X)$. Then the moment generating function for Y is given by

$$M_Y(t) = E_Y[e^{tY}] = E_X\left[e^{t(-2\theta\ln X)}\right] = E[X^{-2\theta t}]$$
$$= \int_0^1 x^{-2\theta t}\theta x^{\theta-1}dx = \int_0^1 \theta x^{(1-2t)\theta-1}dx$$

$$\Rightarrow M_Y(t) = \frac{x^{\theta(1-2t)}}{1-2t}\Big|_{x=0}^{x=1}, \quad \text{provided } t < \frac{1}{2}.$$

Thus, we have

$$M_Y(t) = (1 - 2t)^{-1}, \quad \text{for } t < \frac{1}{2},$$

which we recognize as the moment generating function for a gamma distribution with parameters $\alpha = 1$ and $\beta = 2$. Hence, $Y \sim Gamma\ (\alpha = 1, \beta = 2)$.

Example 6.8 Chi-Square Distribution

Let $X \sim n\ (0, 1)$ and set $Y = X^2$. Then the m.g.f. for Y is given by

$$M_Y(t) = E_Y\left[e^{tY}\right] = E_X\left[e^{tX^2}\right]$$

$$= \int_{-\infty}^{\infty} e^{tx^2} \frac{1}{\sqrt{2\pi}} e^{-\frac{x^2}{2}} dx = \int_{-\infty}^{\infty} \frac{1}{\sqrt{2\pi}} e^{-x^2\left(\frac{1}{2}-t\right)} dx$$

$$= \int_{-\infty}^{\infty} \frac{1}{\sqrt{2\pi}} e^{-x^2/\left[\frac{2}{1-2t}\right]} dx$$

Multiplying the right-hand side of this equation by the "special 1" $\sqrt{\frac{1}{1-2t}}/\sqrt{\frac{1}{1-2t}}$, we see that

$$M_Y(t) = \sqrt{\frac{1}{1-2t}} \int_{-\infty}^{\infty} \frac{1}{\sqrt{2\pi\left[\frac{1}{1-2t}\right]}} e^{-\frac{x^2}{2\left[\frac{1}{1-2t}\right]}} dx$$

But, the integrand in this integral is nothing more than the p.d.f. for the $n\left(0, \frac{1}{1-2t}\right)$ distribution, provided that $t < \frac{1}{2}$, so that the integral itself is equal to 1. Thus, we have

$$M_Y(t) = (1 - 2t)^{-\frac{1}{2}}, \quad \text{provided that } t < \frac{1}{2}.$$

Since this is the moment generating function for the chi-square distribution with one degree of freedom, it follows that $Y = X^2 \sim \chi^2(1)$.

We want to emphasize a special feature of the previous example. We used a "special 1," namely, $1 = \sqrt{\frac{1}{1-2t}}/\sqrt{\frac{1}{1-2t}}$, to "trick" the integrand in question into something that we recognized, namely, the p.d.f. for the $n\left(0, \frac{1}{1-2t}\right)$ distribution, so that the associated integral was automatically known to be equal to 1. This approach enables us to avoid any additional effort in evaluating that integral, and it helps us become more familiar with many of the properties of the most common probability distributions. You should keep this idea in mind throughout this text, as you strive to turn this "special 1" trick into a useful technique. If you prefer, you can continue to work on evaluating difficult integrals (or sums) directly!

6.3   Distribution Function Technique

Let X be a random variable with support space S and set $Y = u(X)$. Then the support space for Y is $T = u(S)$ and the c.d.f. for Y is given by

$$G_Y(y) = P_Y(Y \leq y) = P_X(u(X) \leq y), \quad \text{for } y \in T = u(S).$$

If we recognize this c.d.f., then we can identify the probability distribution for Y. If both X and Y are continuous variables, we can also obtain the p.d.f. for Y by differentiating the c.d.f. $G_Y(y)$. The distribution function technique is the least useful of the three techniques considered in this chapter, but there are situations where it is preferred to the other two.

Example 6.9 Gamma Distribution
Let X be a continuous random variable with p.d.f.:

$$f_X(x) = \theta x^{\theta-1} I_{(0,1)}(x), \quad \text{with } 0 < \theta < \infty.$$

Set $Y = -2\theta \ln(X)$. Then the c.d.f. for Y is given by

$$G_Y(y) = P_Y(Y \leq y) = 0, \quad \text{for } y \leq 0$$

and

$$G_Y(y) = P_X(-2\theta \ln(X) \leq y) = P_X(\ln X \geq -\frac{y}{2\theta})$$

$$= P_X(X \geq e^{-\frac{y}{2\theta}}) = 1 - F_X(e^{-\frac{y}{2\theta}}), \quad \text{for } y > 0,$$

where $F_X(x)$ is the c.d.f. for X. Since X and Y are continuous variables, we can differentiate both sides of this expression with respect to y to obtain the p.d.f. for Y, namely,

$$g_Y(y) = 0, \quad \text{for } y \leq 0,$$

and

$$g_Y(y) = \frac{dG_Y(y)}{dy} = \frac{d\left\{1 - F_X\left(e^{-\frac{y}{2\theta}}\right)\right\}}{dy}, \quad \text{for } 0 < y < \infty.$$

Using the composite function rule for derivatives, it follows that

$$g_Y(y) = -f_X\left(e^{-\frac{y}{2\theta}}\right)\left[e^{-\frac{y}{2\theta}}\left\{-\frac{1}{2\theta}\right\}\right]$$

$$= \frac{1}{2\theta}\left\{\theta\left\{e^{-\frac{y}{2\theta}}\right\}^{\theta-1} e^{-\frac{y}{2\theta}}\right\} = \frac{1}{2}e^{-\frac{y}{2}}, \quad \text{for } 0 < y < \infty.$$

Hence, the p.d.f. for Y is given by

$$g_Y(y) = \frac{1}{2}e^{-\frac{y}{2}}I_{(0,\infty)}(y),$$

which we recognize as the p.d.f. for a gamma distribution with $\alpha = 1$ and $\beta = 2$. Thus, $Y \sim Gamma$ ($\alpha = 1$, $\beta = 2$) or, equivalently, $Y \sim \chi^2(2)$.

Example 6.10 Chi-Square Distribution
Let $X \sim n$ $(0, 1)$ and set $Y = X^2$. Then the c.d.f. for Y is given by

$$G_Y(y) = P_Y(Y \le y) = P_X(X^2 \le y)$$
$$= P_X(-\sqrt{y} \le X \le \sqrt{y}) = \Phi(\sqrt{y}) - \Phi(-\sqrt{y}),$$

where $\Phi(\cdot)$ is the c.d.f. for the standard normal distribution. Since X and Y are continuous variables, we can use the composite function rule for differentiation to see (for $y > 0$) that the p.d.f. for Y is given by

$$g_Y(y) = \frac{dG_Y(y)}{dy} = \frac{d}{dy}[\Phi(\sqrt{y}) - \Phi(-\sqrt{y})] = \frac{\phi(\sqrt{y})}{2\sqrt{y}} + \frac{\phi(-\sqrt{y})}{2\sqrt{y}}$$

where $\phi(\cdot)$ is the standard normal p.d.f. Since the standard normal p.d.f. is symmetric about 0, it follows that

$$g_Y(y) = 2\left[\frac{1}{\sqrt{2\pi}}e^{-\frac{(\sqrt{y})^2}{2}}\right]\frac{1}{2\sqrt{y}} = \frac{1}{\sqrt{2}\sqrt{\pi}}y^{\frac{1}{2}-1}e^{-\frac{y}{2}}.$$

Reminding ourselves that $\sqrt{\pi} = \Gamma\left(\frac{1}{2}\right)$, it follows that

$$g_Y(y) = \frac{1}{2^{\frac{1}{2}}\Gamma\left(\frac{1}{2}\right)}y^{\frac{1}{2}-1}e^{-\frac{y}{2}}I_{(0,\infty)}(y),$$

which is the p.d.f. for a chi-square distribution with 1 degree of freedom. Thus, $Y = X^2 \sim \chi^2(1)$. (We previously established this same result in Example 6.8 using the moment generating function technique. Do you have a preference?)

Example 6.11 All Continuous Roads Lead to the *Unif* (0, 1) Distribution
Let X be a continuous random variable with strictly monotone c.d.f. $F_X(x)$ for $0 < F_X(x) < 1$, and define the new random variable $Y = u(X) = F_X(X)$. Then the c.d.f. for Y is given by

$$G_Y(y) = P_Y(Y \le y) = P_X(F_X(X) \le y)$$
$$= P_X\left(X \le F_X^{-1}(y)\right) = F_X\left(F_X^{-1}(y)\right) = y, \quad \text{for } 0 < y < 1,$$

with $G_Y(y) = 0$, for $y \le 0$, and $G_Y(y) = 1$, for $y \ge 1$. But this is just the p.d.f. for the $Unif\,(0, 1)$ random variable. Thus, $Y = F_X(X) \sim Unif\,(0, 1)$. (The result also holds even if $F_X(x)$ is not strictly monotone for $0 < F_X(x) < 1$ with an appropriate definition for the inverse function $F_X^{-1}(x)$.)

This rather remarkable result is called the **Probability Integral Transformation (PIT theorem)**. It states that for *any* continuous distribution, there is a single transformation $Y = F_X(X)$ that maps the underlying X probability distribution to the same $Unif\,(0, 1)$ distribution. Thus, all continuous probability distributions are linked through this $Unif\,(0, 1)$ connection. A portion of an entire field of statistics, called nonparametric statistics, owes its existence to this simple fact.

Example 6.12 Logistic Distribution
Consider the continuous random variable X with p.d.f.:

$$f_X(x) = \frac{e^{-x}}{(1 + e^{-x})^2} I_{(-\infty,\infty)}(x).$$

(This is known as the *logistic distribution*.) The associated c.d.f. for X is given by

$$F_X(x) = (1 + e^{-x})^{-1}, \quad -\infty < x < \infty.$$

It then follows from the PIT theorem that $Y = (1 + e^{-X})^{-1} = \frac{e^X}{1+e^X}$ has a $Unif\,(0, 1)$ distribution.

6.4 Exercises

6.1. Let $X \sim n(\mu, \sigma^2)$. Use the moment generating function technique to identify the distribution of $Y = aX + b$, for arbitrary constants a and b.

6.2. Let $X \sim Gamma\,(r/2, \beta)$. Use the moment generating function technique to identify the distribution of $Y = 2X/\beta$.

6.3. Let $X \sim n(\mu, \sigma^2)$. Argue that $Y = \left(\frac{X-\mu}{\sigma}\right)^2 \sim \chi^2(1)$.

6.4. Let X be a continuous random variable with p.d.f.:

$$f_X(x) = 2x I_{(0,1)}(x).$$

Find a function $Y = u(X)$ that has a $Unif\,(0, 1)$ distribution.

6.5. Let X be a continuous random variable with p.d.f.:

$$f_X(x) = xe^{-x} I_{(0,\infty)}(x).$$

Find a function $Y = u(X)$ that has a $Unif\,(0, 1)$ distribution.

6.6. Let $X \sim n\,(\mu, \sigma^2)$. Find a function $Y = u(X)$ that has a $Unif\,(0, 1)$ distribution.

6.7. Let X be a continuous random variable with p.d.f.:

$$f_X(x) = \frac{1}{x\sqrt{2\pi\sigma^2}} e^{-(\ln(x)-\mu)^2/2\sigma^2} I_{(0,\infty)}(x).$$

Find the p.d.f. for the random variable $Y = \ln(X)$.

6.8. Let X be a continuous random variable with p.d.f.:

$$f_X(x) = e^{-(x-\theta)} I_{(\theta,\infty)}(x) I_{(-\infty,\infty)}(\theta).$$

Find the p.d.f. for the random variable $Y = X - \theta$.

6.9. Let X be a continuous random variable with p.d.f.:

$$f_X(x) = \frac{1}{6\beta^4} x^3 e^{-\frac{x}{\beta}} I_{(0,\infty)}(x) I_{(0,\infty)}(\beta).$$

Find the p.d.f. for the random variable $Y = \sqrt{X}$.

6.10. Let $X \sim n\,(\mu, \sigma^2)$.

(a) Find the p.d.f. for the random variable $Y = e^X$.
(b) Use the moment generating function for X to find $E[Y]$ and $Var(Y)$.

6.11. Let X be a continuous random variable with p.d.f.:

$$f_X(x) = \frac{1}{2} e^{-|x|} I_{(-\infty,\infty)}(x).$$

Find and identify the p.d.f. for $Y = |X|$.

6.12. Let Y be a continuous random variable with p.d.f.:

$$f_Y(y) = \frac{y^7}{3\beta^4} e^{-\frac{y^2}{\beta}} I_{(0,\infty)}(y) I_{(0,\infty)}(\beta).$$

Find the p.d.f. for the random variable $Q = \frac{Y^2}{\beta}$.

6.13. Let Y be a continuous random variable with p.d.f.:

$$f_Y(y) = \frac{\Gamma(\alpha+\beta)}{\Gamma(\alpha)\Gamma(\beta)} y^{\alpha-1} (1-y)^{\beta-1} \ I_{(0,1)}(y) \ I_{(0,\infty)}(\infty) I_{(0,\infty)}(\beta).$$

(a) Identify the probability distribution of Y.
(b) Find the p.d.f. for the random variable $Q = 1 - Y$, and identify the distribution.

6.14. Let U be a continuous random variable with p.d.f.:

$$f_U(u) = 6u(1-u)I_{(0,1)}(u).$$

(a) Identify the probability distribution of U.
(b) Find the p.d.f. for the random variable $W = $ maximum $\{U, 1 - U\}$.

6.15. Let X be a continuous random variable with p.d.f.:

$$f_X(x) = 5(1-x)^4 \ I_{(0,1)}(x).$$

Find the p.d.f. for the random variable $Y = 1 - (1 - X)^5$.

6.16. Let X be a continuous random variable with p.d.f.:

$$f_X(x) = 7x^6 \ I_{(0,1)}(x).$$

Find the p.d.f for the random variable $Y = X^7$ and identify the distribution.

6.17. Let $X \sim Exp\,(\lambda)$. Find the p.d.f. for the variable $W = \sqrt{X}$.

6.18. Let $X \sim Poisson\,(\lambda)$, with $\lambda > 0$. Find the p.d.f. for the random variable $W = 2X^2$.

6.19. Let X be a continuous random variable with p.d.f.:

$$f_X(x) = 2xe^{-x^2} I_{(0,\infty)}(x).$$

Find the p.d.f. for the random variable $Y = X^2$.

6.20. Let X be a discrete random variable with p.d.f.:

$$f_X(x) = \left(\frac{1}{2}\right)^x I_{\{1,2,3,...\}}(x).$$

Find the p.d.f. for the random variable $Y = X^3$.

6.21. Let X be a continuous random variable with p.d.f.:

$$f_X(x) = \frac{1}{\pi} I_{\left(-\frac{\pi}{2}, \frac{\pi}{2}\right)}(x).$$

Find the p.d.f. for the random variable $Y = \tan(X)$. (This is known as the *Cauchy distribution*.)

6.22. Let $X \sim Unif(0, 1)$. Find the p.d.f. for $Y = -\ln X$ and identify the distribution.

6.23. Let $X \sim Unif(-2, 2)$. Find the p.d.f. for the random variable $Y = X^2$.

6.24. Let $X \sim Unif(-2, 1)$. Find the p.d.f. for the random variable $Y = X^2$. [Hint: Consider $\{-2 < x < -1\}$ and $\{-1 < x < 1\}$ separately.]

6.25. Let $X \sim Exp(1)$ and set $Y = \frac{e^{-X}}{1+e^{-X}}$. Find the p.d.f. for Y.

6.26. Let $X \sim Exp(\lambda)$ and set $Y = e^{-X}$. Find the p.d.f. for Y.

6.27. Let $X \sim Unif(0, \theta)$, with $\theta > 0$. Find the p.d.f. for $W = e^X$.

6.28. Let X be a continuous random variable with p.d.f.:

$$f_X(x) = \theta\, x^{\theta-1} I_{(0,1)}(x) I_{(0,\infty)}(\theta).$$

(a) Find the p.d.f. for $Y = -(\ln X)^{-1}$ and evaluate $P\left(\frac{Y}{2} < \theta < Y\right)$.
(b) Let $W = -2\theta \ln X$. Show that the probability distribution for W does not depend on θ.

6.29. Let A be the area of a circle with radius $R \sim Exp(\beta)$. Find the p.d.f. for A.

6.30. Let $X \sim Gamma(4, 2)$. Find a function $u(X)$ that has a $Unif(0, 1)$ distribution.

6.31. Let X be a continuous random variable with p.d.f.:

$$f_X(x) = 4x^3 I_{(0,1)}(x).$$

Find the p.d.f. for $W = -4 \ln(X)$.

6.32. Let $Y \sim NegBin(m, p)$. Find the p.d.f. for $V = Y - m$. Give an interpretation for the random variable V.

6.33. Let X be a continuous random variable with p.d.f.:

$$f_X(x) = \frac{3}{2x^2} I_{(1,3)}(x).$$

Find the p.d.f. for $V = X^2$.

6.34. Let X have a hypergeometric distribution with parameters N, b, and n. Let $Q = n - X$. Find the p.d.f. for Q, being sure to clearly describe its support space. What does Q represent?

6.35. Let X be a discrete random variable with a uniform distribution on the set of integers $S = \{-n, -n + 1, \ldots, -1, 0, 1, \ldots, n - 1, n\}$. Find the p.d.f. for the random variable $Y = X^2$.

6.36. Let X be a continuous random variable with p.d.f.:

$$f_X(x) = \frac{1}{x \ln 2} I_{(1,2)}(x).$$

Find the p.d.f. for the random variable $Y = \ln X$.

6.37. Let X be a continuous random variable with p.d.f.:

$$f_X(x) = \frac{1}{x \ln 2} I_{(1,2)}(x).$$

Find the p.d.f. for the random variable $Y = e^X$.

6.38. You play a game where you roll a pair of fair six-sided dice. Let Y be the sum of the two numbers on the dice, and suppose you win $2Y$ dollars once the roll is complete. Find the p.d.f. for the random amount of your winning. How much should you be asked to play this game to make it fair?

6.39. Let $X \sim Unif(0, 1)$ and let $a < b$ be constants. Consider the linear transformation $Y = a + bX$.

(a) What choices of a and b will lead to $Y \sim Unif(a, a + b)$?
(b) What choices of a and b will lead to $Y \sim Unif(b, a + b)$?
(c) What choices of a and b will lead to $Y \sim Unif(a, b)$?
(d) What choices of a and b will lead to $Y \sim Unif(a, 2a)$?
(e) What choices of a and b will lead to $Y \sim Unif(a, 2b)$?

6.40. Let X be a random variable with mean μ and variance σ^2 and consider the transformation $Y = \frac{X-\mu}{\sigma}$.

(a) Show that $E[Y] = 0$ and $Var(Y) = 1$. (This transformation is often referred to as the *standardization transformation*. Why do you think it has been given this name?)
(b) Suppose $X \sim n(\mu, \sigma^2)$. What is the distribution of Y for this setting?

6.41. Suppose that $h(-x) = - h(x)$ and $k(-x) = k(x)$ for all real x. [We call $h(x)$ an *odd function* and $k(x)$ an *even function* of x.] If all the integrals exist, show that

$$\int_{-\infty}^{\infty} h(x)dx = 0$$

and

$$\int_{-\infty}^{\infty} k(x)dx = 2 \int_{0}^{\infty} k(x)dx.$$

6.42. Let X be a continuous random variable with p.d.f.:

$$f_X(x) = \frac{1}{2} e^{-|x|} I_{(-\infty,\infty)}(x).$$

(This is called the *standard double exponential* or *Laplace* distribution.) Use the results in Exercise 6.41 to find $E[X]$ and $Var(X)$ directly, without first making the transformation $Y = |X|$.

6.43. Let X be a continuous random variable with p.d.f.:

$$f_X(x) = \frac{1}{2\sigma} e^{-|x-\mu|/\sigma} I_{(-\infty,\infty)}(x) I_{(-\infty,\infty)}(\mu) I_{(0,\infty)}(\sigma).$$

(This is the most general form of the *double exponential* or *Laplace* distribution.) Use the results in Exercises 6.40 and 6.42 to show that $E[X] = \mu$ and $Var(X) = 2\sigma^2$.

6.44. We have seen in Exercises 6.40–6.43 that standardizing a random variable by subtracting off its mean and dividing by its standard deviation can lead to simpler calculations in some settings—but is this universally the case? Let $X \sim Poisson\ (\lambda)$. Then we know that $E[X] = Var(X) = \lambda$. Find the p.d.f. for the standardized variable $W = \frac{X-\lambda}{\sqrt{\lambda}}$. Does this make calculations involving the Poisson distribution easier? Why or why not?

6.45. Let $X \sim Exp\ (\beta)$ and set $Y = X^{1/\gamma}$, where $\gamma > 0$.

(a) Find the p.d.f. for Y and verify that it is, indeed a p.d.f. (This is called the *Weibull distribution with parameters γ and β*.)
(b) Find $E[Y]$ and $Var(Y)$.

6.46. Let $X \sim Exp\ (\beta)$ and set $Y = (2X/\beta)^{1/2}$.

(a) Find the p.d.f. for Y and verify that it is, indeed, a p.d.f. (This is called the *Rayleigh distribution with parameter β*.)
(b) Find $E[Y]$ and $Var(Y)$.

6.47. Let $X \sim Gamma\,(\alpha, \beta)$ and set $Y = 1/X$.

(a) Find the p.d.f. for Y and verify that it is, indeed, a p.d.f. (This is called the *Inverted Gamma distribution with parameters α and β.*)

(b) Find $E[Y]$ and $Var(Y)$.

6.48. Let $X \sim Gamma\left(\frac{3}{2}, \beta\right)$ and set $Y = (X/\beta)^{1/2}$.

(a) Find the p.d.f. for Y and verify that it is, indeed, a p.d.f. (This is called the *Maxwell distribution with parameter β.*)

(b) Find $E[Y]$ and $Var(Y)$.

6.49. Let $X \sim Exp\,(1)$ and set $Y = \alpha - \lambda \ln(X)$, where $-\infty < \alpha < \infty$ and $\lambda > 0$.

(a) Find the p.d.f. for Y and verify that it is, indeed, a p.d.f. (This is called the *Gumbel or extreme value distribution with parameters α and λ.*)

(b) Find $E[Y]$ and $Var(Y)$.

6.50. Let X be a continuous random variable with p.d.f.:

$$f_X(x) = \frac{2}{\sqrt{2\pi}}\, e^{-x^2/2}\, I_{(0,\infty)}(x).$$

(This is known as the *folded normal distribution.*)

(a) Show that $f_X(x)$ is, indeed, a p.d.f.

(b) Find $E[X]$ and $Var(X)$.

(c) Find a transformation $Y = u(X)$ and values of $\alpha > 0$ and $\beta > 0$ so that $Y \sim Gamma\,(\alpha, \beta)$.

6.51. Let X be a continuous random variable with p.d.f.:

$$f_X(x) = \frac{3}{8}(x+1)^2 I_{(-1,1)}(x).$$

Find the p.d.f. for $Y = 1 - X^2$ and show that it integrates to 1.

Chapter 7
Sampling Distributions

In Chap. 3, we introduced the concept of a random variable and its probability distribution and devoted Chap. 4 to developing some of the important features of probability distributions. In Chap. 5, we extended these ideas to probability distributions for two variables and in Chap. 6, we discussed how to obtain probability distributions for functions of a single random variable.

In settings where statistics plays an important role in the analysis of and proper interpretation of data obtained from a scientific study, we are, however, almost always dealing with much more than the observed outcome for a single (or even two) random variable(s). Good scientific studies involve either replications of a designed experiment or multiple observations obtained from a population of interest. Analyzing data of this type requires that we understand the proper process for combining information from such replications or multiple observations to reach meaningful statistical conclusions about the experiment or population of interest. This leads directly to the all-important concepts of random samples and sampling distributions, the topics of this chapter.

7.1 Simple Random Samples

Definition 7.1 Let $X_1, \ldots, X_n$ be independent and identically distributed (we use the notation *i.i.d.* to represent this) random variables with common p.d.f. $f_X(x)$ and c.d.f. $F_X(x)$. We refer to $X_1, \ldots, X_n$ as a **random sample of size n from** $f_X(x)$. The joint p.d.f. for $X_1, \ldots, X_n$ is given by

$$f_{X_1, \ldots, X_n}(x_1, \ldots, x_n) = \prod_{i=1}^{n} f_X(x_i),$$

(continued)

© Springer Nature Switzerland AG 2020
D. Wolfe, G. Schneider, *Primer for Data Analytics and Graduate Study in Statistics*,
https://doi.org/10.1007/978-3-030-47479-9_7

Definition 7.1 (continued)
and the joint c.d.f. is

$$F_{X_1,\ldots,X_n}(x_1, \ldots, x_n) = P(X_1 \le x_1, \ldots, X_n \le x_n) = \prod_{i=1}^{n} P(X_i \le x_i)$$

$$= \prod_{i=1}^{n} F_{X_i}(x_i).$$

We often refer to the joint p.d.f. as the **likelihood function, L, for the random sample.**

Example 7.1 Normal Distribution

Let $X_1,\ldots,X_n$ be a random sample from the $n\,(\mu, \sigma^2)$ distribution. Then the likelihood function for this random sample is

$$\text{Likelihood Function} = L = \prod_{i=1}^{n} \frac{1}{\sqrt{2\pi\sigma^2}} e^{-\frac{1}{2\sigma^2}(x_i-\mu)^2} = \left(2\pi\sigma^2\right)^{-\frac{n}{2}} e^{-\frac{1}{2\sigma^2}\sum_{i=1}^{n}(x_i-\mu)^2}.$$

Example 7.2 Bernoulli Distribution

Let $X_1,\ldots,X_n$ be a random sample from the *Bernoulli* (θ) distribution, with $0 \le \theta \le 1$. Then the likelihood function for this random sample is

$$L = \prod_{i=1}^{n} \theta^{x_i}(1-\theta)^{1-x_i} I_{\{0,1\}}(x_i) = \theta^{\sum_{i=1}^{n} x_i}(1-\theta)^{n-\sum_{i=1}^{n} x_i} \prod_{i=1}^{n} I_{\{0,1\}}(x_i).$$

Example 7.3 Discrete Distribution

Let X_1, X_2, X_3 be a random sample of size $n = 3$ from the discrete distribution with p.d.f.

$$f_X(x) = \frac{1}{4}, x = -1, +1,$$

$$= \frac{1}{2}, x = 0,$$

$$= 0, \text{elsewhere.}$$

The associated likelihood function for this sample is given by

$$L = \prod_{i=1}^{3} f_X(x_i)$$

$$= \frac{1}{64}, (x_1, x_2, x_3) \ni x_i = \pm 1, i = 1, 2, 3,$$

$$= \frac{1}{32}, (x_1, x_2, x_3) \ni \text{two of the } x_i\text{'s are } \pm 1 \text{ and one of the } x_i\text{'s is } 0,$$

$$= \frac{1}{16}, (x_1, x_2, x_3) \ni \text{one of the } x_i\text{'s is } \pm 1 \text{ and two of the } x_i\text{'s are } 0,$$

$$= \frac{1}{8}, (x_1, x_2, x_3) = (0, 0, 0),$$

$$= 0, \text{elsewhere.}$$

Note that this likelihood function can be written in a more concise form as follows:

$$L = \frac{3!}{N_1! N_2! N_3!} \left(\frac{1}{2}\right)^{N_1} \left(\frac{1}{4}\right)^{N_2} \left(\frac{1}{4}\right)^{N_3} \prod_{i=1}^{3} I_{\{0,1,2,3\}}(N_i) \ni N_1 + N_2 + N_3 = 3,$$

where $N_1 = $ (number of X's equal to 0), $N_2 = $ (number of X's equal to $- 1$), and $N_3 = $ (number of X's equal to $+ 1$).

Definition 7.2 Let $X_1, \ldots, X_n$ be a random sample of size n from a probability distribution with p.d.f. $f_X(x)$. Any random variable $u(X_1, \ldots, X_n)$ that is a function of the sample data $X_1, \ldots, X_n$ only is called a **statistic**.

Example 7.4 Some Useful Statistics

$$\bar{X} = \frac{1}{n} \sum_{i=1}^{n} X_i \ (\text{sample mean})$$

$$S^2 = \frac{1}{n} \sum_{i=1}^{n} (X_i - \bar{X})^2 = \frac{1}{n} \sum_{i=1}^{n} X_i^2 - \bar{X}^2 \ (\text{sample variance})$$

$$X_{(n)} = \text{maximum } \{X_1, \ldots, X_n\} \ (\text{sample maximum})$$

$$X_{(1)} = \text{minimum } \{X_1, \ldots, X_n\} \ (\text{sample minimum})$$

$$X_{(1)} \leq X_{(2)} \leq \cdots \leq X_{(n)} \ (\text{sample order statistics})$$

$$\tilde{X} = \text{median}\{X_1, \ldots, X_n\}(\text{sample median})$$

$$\text{Other Statistics:} \qquad \prod_{i=1}^{n} X_i \qquad -\sum_{i=1}^{n} \ln(X_i) \qquad \frac{X_{(1)} + X_{(n)}}{2}$$

$$\sum_{i=1}^{n} |X_i - \overline{X}| \qquad \sum_{i=1}^{n} iX_i \qquad \sum_{i=1}^{n-1} (X_{i+1} + X_i)^2.$$

7.2   Sampling Distributions

Definition 7.3 Let $X_1, \ldots, X_n$ be a random sample, and let $T = T(X_1, \ldots, X_n)$ be a statistic based on this sample. Then, T is itself a random variable, and it therefore has a probability distribution of its own. We refer to the probability distribution of the statistic T as its **sampling distribution**. The sampling distribution of T has all the usual properties of any probability distribution, such as mean μ_T, variance σ_T^2, p.d.f. $f_T(t)$, c.d.f. $F_T(t)$, m.g.f $M_T(t)$, etc.

Example 7.5 Discrete Distribution
Consider a population consisting of the five values $\{6, 6, 10, 18, 20\}$. Suppose we select three observations from this population at random and let X_i represent the outcome of the ith number selected, $i = 1, 2, 3$. We want to find the sampling distribution of the two statistics:

$$\widetilde{X} = \text{median } \{X_1, X_2, X_3\} = X_{(2)} \text{ and } \overline{X} = \frac{1}{3}\sum_{i=1}^{3} X_i = \text{sample mean.}$$

First, note that:

$$E[X_1] = E[X_2] = E[X_3] = \frac{1}{5}[6 + 6 + 10 + 18 + 20] = \frac{60}{5} = 12$$

and

$$Var(X_1) = Var(X_2) = Var(X_3)$$
$$= \frac{1}{5}\left[(6-12)^2 + (6-12)^2 + (10-12)^2 + (18-12)^2 + (20-12)^2\right] = 35.2.$$

Setting 1: Sampling with Replacement Between Selections
In this setting, X_1, X_2, X_3 is a random sample of size $n = 3$ from the probability distribution with p.d.f.

$$f_X(x) = \frac{1}{5}, x = 10, 18, 20$$
$$= \frac{2}{5}, x = 6$$
$$= 0, \text{elsewhere.}$$

We obtain the sampling distributions for $\widetilde{X}$ and $\overline{X}$ directly by enumeration, as follows:

Outcome X_1, X_2, X_3	Value of $\widetilde{X}$	Value of $\overline{X}$	Probability
(6, 6, 6)	6	6	$\left(\frac{2}{5}\right)^3 = \frac{8}{125}$
(6, 6, 10)	6	$7\frac{1}{3}$	$\binom{3}{1}\left(\frac{2}{5}\right)^2\left(\frac{1}{5}\right) = \frac{12}{125}$
(6, 6, 18)	6	10	$\binom{3}{1}\left(\frac{2}{5}\right)^2\left(\frac{1}{5}\right) = \frac{12}{125}$
(6, 6, 20)	6	$10\frac{2}{3}$	$\binom{3}{1}\left(\frac{2}{5}\right)^2\left(\frac{1}{5}\right) = \frac{12}{125}$
(6, 10, 10)	10	$8\frac{2}{3}$	$\binom{3}{1}\left(\frac{1}{5}\right)^2\left(\frac{2}{5}\right) = \frac{6}{125}$
(6, 10, 18)	10	$11\frac{1}{3}$	$3!\left(\frac{1}{5}\right)^2\left(\frac{2}{5}\right) = \frac{12}{125}$
(6, 10, 20)	10	12	$3!\left(\frac{1}{5}\right)^2\left(\frac{2}{5}\right) = \frac{12}{125}$
(6, 18, 18)	18	14	$\binom{3}{1}\left(\frac{1}{5}\right)^2\left(\frac{2}{5}\right) = \frac{6}{125}$
(6, 18, 20)	18	$14\frac{2}{3}$	$3!\left(\frac{1}{5}\right)^2\left(\frac{2}{5}\right) = \frac{12}{125}$
(6, 20, 20)	20	$15\frac{1}{3}$	$\binom{3}{1}\left(\frac{1}{5}\right)^2\left(\frac{2}{5}\right) = \frac{6}{125}$
(10, 10, 10)	10	10	$\left(\frac{1}{5}\right)^3 = \frac{1}{125}$
(10, 10, 18)	10	$12\frac{2}{3}$	$\binom{3}{1}\left(\frac{1}{5}\right)^3 = \frac{3}{125}$
(10, 10, 20)	10	$13\frac{1}{3}$	$\binom{3}{1}\left(\frac{1}{5}\right)^3 = \frac{3}{125}$
(10, 18, 18)	18	$15\frac{1}{3}$	$\binom{3}{1}\left(\frac{1}{5}\right)^3 = \frac{3}{125}$
(10, 18, 20)	18	16	$3!\left(\frac{1}{5}\right)^3 = \frac{6}{125}$

Outcome X_1, X_2, X_3	Value of $\tilde{X}$	Value of $\overline{X}$	Probability
(10, 20, 20)	20	$16\frac{2}{3}$	$\binom{3}{1}\left(\frac{1}{5}\right)^3 = \frac{3}{125}$
(18, 18, 18)	18	18	$\left(\frac{1}{5}\right)^3 = \frac{1}{125}$
(18, 18, 20)	18	$18\frac{2}{3}$	$\binom{3}{1}\left(\frac{1}{5}\right)^3 = \frac{3}{125}$
(18, 20, 20)	20	$19\frac{1}{3}$	$\binom{3}{1}\left(\frac{1}{5}\right)^3 = \frac{3}{125}$
(20, 20, 20)	20	20	$\left(\frac{1}{5}\right)^3 = \frac{1}{125}$

Thus, under sampling with replacement between selections, the sampling distributions of $\tilde{X}$ and $\overline{X}$ are, respectively, given by

$$P(\tilde{X} = 6) = \frac{44}{125}, \quad P(\tilde{X} = 10) = \frac{37}{125}, \quad P(\tilde{X} = 18) = \frac{31}{125}, \quad \text{and} \quad P(\tilde{X} = 20) = \frac{13}{125}$$

and

$$P(\overline{X} = t) = \frac{8}{125}, \quad t = 6$$
$$= \frac{12}{125}, \quad t = 7\frac{1}{3}, 10\frac{2}{3}, 11\frac{1}{3}, 12, 14\frac{2}{3}$$
$$= \frac{13}{125}, \quad t = 10$$
$$= \frac{6}{125}, \quad t = 8\frac{2}{3}, 14, 16$$
$$= \frac{3}{125}, \quad t = 12\frac{2}{3}, 13\frac{1}{3}, 16\frac{2}{3}, 18\frac{2}{3}, 19\frac{1}{3}$$
$$= \frac{9}{125}, \quad t = 15\frac{1}{3}$$
$$= \frac{1}{125}, \quad t = 18, 20$$
$$= 0, \quad \text{elsewhere.}$$

Now, we can obtain the means for these two sampling distributions with replacement between selections, as follows:

$$E[\tilde{X}] = \mu_{\tilde{X}} = \sum_x \tilde{x} P(\tilde{X} = \tilde{x}) = 6\left(\frac{44}{125}\right) + 10\left(\frac{37}{125}\right) + 18\left(\frac{31}{125}\right) + 20\left(\frac{13}{125}\right)$$

$$= 11.616$$

and

$$E[\overline{X}] = \mu_{\overline{X}} = \sum_{\overline{x}} \overline{x}\, P(\overline{X} = \overline{x}) = \left\{ 6\left(\frac{8}{125}\right) + \left[7\frac{1}{3} + 10\frac{2}{3} + 11\frac{1}{3} + 12 + 14\frac{2}{3}\right]\left(\frac{12}{125}\right) \right.$$

$$+ 10\left(\frac{13}{125}\right) + \left[8\frac{2}{3} + 14 + 16\right]\left(\frac{6}{125}\right)$$

$$+ \left[12\frac{2}{3} + 13\frac{1}{3} + 16\frac{2}{3} + 18\frac{2}{3} + 19\frac{1}{3}\right]\left(\frac{3}{125}\right)$$

$$\left. + 15\frac{1}{3}\left(\frac{9}{125}\right) + [18 + 20]\left(\frac{1}{125}\right) \right\}$$

$$= \frac{48}{125} + \frac{672}{125} + \frac{130}{125} + \frac{232}{125} + \frac{242}{125} + \frac{138}{125} + \frac{38}{125}$$

$$= \frac{1500}{125} = 12.$$

Think About It Notice that $\mu_{\overline{X}} = \mu_X = 12$. Is this surprising?

Think About It For Setting 1 in Example 7.5, we found that $E[\widetilde{X}] = 11.616 < E[\overline{X}] = 12$. Why is this inequality not surprising, given the nature of the underlying population?

Setting 2: Sampling Without Replacement Between Selections

In this setting, X_1, X_2, X_3 is not a random sample from the underlying probability distribution—in fact, the X's are dependent random variables. Here, however, all of the $\binom{5}{3} = 10$ possible subsets of three numbers selected from the population $\{6, 6, 10, 18, 20\}$ are equally likely, so that each has probability $\frac{1}{10}$ of being chosen as $\{X_1, X_2, X_3\}$, without regard to order. Thus, the joint probability distribution for the set of three observations $\{X_1, X_2, X_3\}$ (without regard to order) is given by

$$P((X_1, X_2, X_3) = (x_1, x_2, x_3)) = \frac{1}{10}$$

$$\text{for } (x_1, x_2, x_3) \in \{(6, 6, 10), (6, 6, 18), (6, 6, 20), (6, 10, 18),$$

$$(6, 10, 20) \ (6, 18, 20), (6, 10, 18), (6, 10, 20),$$

$$(6, 18, 20), (10, 18, 20)\}.$$

Once again, we obtain the sampling distributions for $\widetilde{X}$ and $\overline{X}$ directly by enumeration, as follows:

Outcome (X_1, X_2, X_3)	Value of $\widetilde{X}$	Value of $\overline{X}$	Probability
(6, 6, 10)	6	$7\frac{1}{3}$	$\frac{1}{10}$
(6, 6, 18)	6	10	$\frac{1}{10}$
(6, 6, 20)	6	$10\frac{2}{3}$	$\frac{1}{10}$
(6, 10, 18)	10	$11\frac{1}{3}$	$\frac{2}{10}$

Outcome (X_1, X_2, X_3)	Value of $\tilde{X}$	Value of $\bar{X}$	Probability
(6, 10, 20)	10	12	$\frac{2}{10}$
(6, 18, 20)	18	$14\frac{2}{3}$	$\frac{2}{10}$
(10, 18, 20)	18	16	$\frac{1}{10}$

Thus, under sampling without replacement between selections, the sampling distributions of $\tilde{X}$ and $\bar{X}$ are, respectively, given by

$$P(\tilde{X} = 6) = \frac{3}{10} \quad P(\tilde{X} = 10) = \frac{4}{10} \quad P(\tilde{X} = 18) = \frac{3}{10}$$

and

$$P\left(\bar{X} = 7\frac{1}{3}\right) = P(\bar{X} = 10) = P\left(\bar{X} = 10\frac{2}{3}\right) = P(\bar{X} = 16) = \frac{1}{10}$$

$$P\left(\bar{X} = 11\frac{1}{3}\right) = P(\bar{X} = 12) = P\left(\bar{X} = 14\frac{2}{3}\right) = \frac{2}{10}.$$

The means for these sampling distributions without replacement between selections are given by

$$E\left[\bar{X}\right] = \mu_{\bar{X}} = \left\{7\frac{1}{3} + 10 + 10\frac{2}{3} + 16\right\} \times \frac{1}{10} + \left\{11\frac{1}{3} + 12 + 14\frac{2}{3}\right\} \times \frac{2}{10} = 12 = \mu_X$$

and

$$E[\tilde{X}] = \left(6 \times \frac{3}{10}\right) + \left(10 \times \frac{4}{10}\right) + \left(18 \times \frac{3}{10}\right) = 11.2.$$

Note that $\mu_{\bar{X}} = \mu_X = 12$ under both sampling with or without replacement between selections.

> **Definition 7.4** Let $X_1, \ldots, X_n$ be a random sample from a probability distribution with mean μ_X, and let $T(X_1, \ldots, X_n)$ be a statistic based on this sample. If $E[T(X_1, .., X_n)] = \mu_X$, we say that $T(X_1, \ldots, X_n)$ is an **unbiased estimator for** μ_X.

Thus, in Example 7.5, the sample mean $\bar{X}$ is an unbiased estimator for μ_X under either sampling with or without replacement between selections. In Exercise 7.6, you are asked to show that this is true for any underlying population for which the mean μ_X exists.

In Exercises 7.4 and 7.5, you are asked to show that $Var(\bar{X}) = 11.733$ in Setting 1 when sampling with replacement between selections and $Var(\bar{X}) = 5.867$ in Setting 2 when sampling without replacement between selections. Note that

$$Var(\bar{X}) \text{ under Setting } 2 \ = 5.867 = \frac{11.733}{2} = \frac{1}{2} Var(\bar{X}) \text{ under Setting } 1.$$

Think About It Does the smaller value for $Var(\bar{X})$ when sampling without replacement between selections make intuitive sense? What would $Var(\bar{X})$ be if we had selected a sample of size $n = 5$ from this population without replacement between selections?

This is a specific example of a more general fact that the variance for the sample mean when sampling with replacement between selections must be reduced when, instead, the sampling is done without replacement between selections. In fact, there is a formula that provides the exact relationship between $Var(\bar{X})$ for these two settings when the underlying population is finite. Let N be the size of the underlying finite population and consider taking a random sample of size n from the population. Then

$$Var(\bar{X}) \text{without replacement} = \left(\frac{N-n}{N-1}\right) \times \left[Var(\bar{X}) \text{with replacement}\right].$$

The factor $\frac{N-n}{N-1}$ is called the *finite population correction factor*. What happens to the finite population correction factor when N is much larger than n? What does this say about $Var(\bar{X})$ for sampling without replacement from very large populations? What happens when n gets closer and closer to the full population size N?

7.3   General Approaches for Obtaining Sampling Distributions

While the previous (long!) Example 7.5 provided a good setting to fully understand the process for moving from a population to the sampling distribution of a statistic, we certainly hope that we do not *always* have to enumerate such details to obtain sampling distributions for statistics of interest. Fortunately, there are statistical techniques that enable us to avoid such enumeration in most settings. Specifically, we now discuss three standard approaches for obtaining sampling distributions that avoid such enumeration: (1) change of variable technique, (2) moment generating function technique, and (3) c.d.f. technique.

Let $X_1, \ldots, X_n$ be a random sample from a probability distribution with p.d.f. $f_X(x)$, m.g.f. $M_X(t)$, and c.d.f. $F_X(x)$. Let $U = U(X_1, \ldots, X_n)$ be a statistic that is of interest. We now describe and illustrate three different approaches for finding the probability distribution of U.

7.3.1   *Moment Generating Function Technique*

Let $M_U(t)$ be the m.g.f. for U. Then we have

$$M_U(t) = E_U\left[e^{tU}\right] = E_{X_1,\ldots,X_n}\left[e^{tU(X_1,\ldots,X_n)}\right]$$

$$= \int_{-\infty}^{\infty}\int_{-\infty}^{\infty}\cdots\int_{-\infty}^{\infty} e^{tU(X_1,\ldots,X_n)} f_X(x_1)\cdots f_X(x_n)dx_1\cdots dx_n$$

(or $\sum_{-\infty}^{\infty}\sum_{-\infty}^{\infty}\cdots\sum_{-\infty}^{\infty} e^{tU(x_1,\ldots,x_n)} f_X(x_1)\cdots f_X(x_n)$ in the discrete setting).

Then, if we can recognize the resulting moment generating function, we can also deduce the associated probability distribution for U.

Most Useful When $U = \sum_{i=1}^{n} X$ or when $U = \ln\left[V(X_1,\ldots,X_n)\right]$ for some "nice" statistic V.

Example 7.6 Sum of Independent Random Variables

Let $X_1,\ldots,X_n$ be independent random variables with moment generating functions $M_{X_1}(t), \ldots, M_{X_n}(t)$, respectively. Then the moment generating function for $U = \sum_{i=1}^{n} X_i$ is given by

$$M_U(t) = E_U\left[e^{tU}\right] = E_{X_1,\ldots,X_n}\left[e^{t\sum_{i=1}^{n}X_i}\right] = E_{X_1,\ldots,X_n}\left[\prod_{i=1}^{n}e^{tX_i}\right] \overset{independent}{=} \prod_{i=1}^{n}E_{X_i}\left[e^{tX_i}\right]$$

$$= \prod_{i=1}^{n} M_{X_i}(t).$$

If, in addition, the X's are identically distributed (so that $X_1,\ldots,X_n$ is a random sample from a distribution) with common moment generating function $M_X(t)$, it follows that

$$M_U(t) = [M_X(t)]^n.$$

Example 7.7 Sum of Independent Normal Variables

Let $X_1,\ldots,X_n$ be independent random variables with $X_i \sim n(\mu_i,\sigma_i^2)$, $i = 1\ldots, n$. Then, from Example 7.6, we see that the moment generating function for $U = \sum_{i=1}^{n} X_i$ is given by

$$M_U(t) = \prod_{i=1}^{n} M_{X_i}(t) = \prod_{i=1}^{n} e^{\mu_i t + \frac{\sigma_i^2}{2}t^2} = e^{\left(\sum_{i=1}^{n}\mu_i\right)t + \left(\sum_{i=1}^{n}\sigma_i^2\right)\frac{t^2}{2}},$$

which is the moment generating function for the $n\left(\sum_{i=1}^{n}\mu_i, \sum_{i=1}^{n}\sigma_i^2\right)$ distribution. It follows that

$$U = \sum_{i=1}^{n} X_i \sim n\left(\sum_{i=1}^{n}\mu_i, \sum_{i=1}^{n}\sigma_i^2\right).$$

If the X's are identically distributed with common mean μ and common variance σ^2, then

$$U \sim n\left(n\mu, n\sigma^2\right),$$

so that $E[U] = n\mu$ and $V(U) = n\sigma^2$, which leads to

$$E\left[\frac{U}{n}\right] = E[\bar{X}] = \frac{n\mu}{n} = \mu \text{ and } Var\left(\frac{U}{n}\right) = Var(\bar{X}) = \frac{n\sigma^2}{n^2} = \frac{\sigma^2}{n}.$$

Example 7.8 Sum of Independent Binomial Variables
Let $X_1,\ldots,X_k$ be independent random variables such that $X_i \sim Binom\ (n_i, p)$, $i = 1,$ $\ldots, k$, with $0 \le p \le 1$. Let $U = \sum_{i=1}^{n} X_i$. Then, from Example 7.6 we see that the moment generating function for U is given by

$$M_U(t) = \prod_{i=1}^{k} M_{X_i}(t) = \prod_{i=1}^{k} [(1-p) + pe^t]^{n_i} = [(1-p) + pe^t]^{\sum_{i=1}^{k} n_i},$$

from which it follows immediately that $U \sim Binom\left(\sum_{i=1}^{k} n_i, p\right)$.

Think About It   Does a similar result hold if the X's do not have a common value of p?

Example 7.9 Sum of Independent Gamma Variables
Let $X_1,\ldots,X_n$ be independent random variables such that $X_i \sim Gamma\ (\alpha_i, \beta)$, with $\beta > 0$ and $\alpha_i > 0$, for $i = 1, \ldots, n$. Let $U = \sum_{i=1}^{n} X_i$. Then the moment generating function for U is given by

$$M_U(t) = \prod_{i=1}^{n} M_{X_i}(t) = \prod_{i=1}^{n} (1-\beta t)^{-\alpha_i} = (1-\beta t)^{-\sum_{i=1}^{n}\alpha_i} \text{ for } t < \frac{1}{\beta}.$$

We recognize this to be the moment generating function for a gamma distribution with parameters $\sum_{i=1}^{n} \alpha_i$ and β. It follows that $U \sim Gamma\left(\sum_{i=1}^{n} \alpha_i, \beta\right)$.

Think About It Does a similar result hold if the X's do not have a common value of β?

7.3.2 Distribution Function Technique

Let $G(u)$ be the c.d.f. for $U = U(X_1, \ldots, X_n)$; that is,

$$G(u) = P_U(U \le u) = P_{X_1, \ldots, X_n}(U(X_1, \ldots, X_n) \le u)$$

$$= \int_{\{U(x_1, \ldots, x_n) \le u\}} \cdots \int f_X(x_1) \cdots f_X(x_n) dx_1 \cdots dx_n$$

(or $\displaystyle\sum_{\{U(x_1, \ldots, x_n) \le u\}} \cdots \sum f_X(x_1) \cdots f_X(x_n)$ in the discrete setting).

Reminder In the continuous setting, we can obtain the p.d.f. $g(u)$ for U directly by differentiating its c.d.f. $G(u)$.

Most Useful
(i) when $n = 1$ and U is a monotone function of X or $U = X^2$ or $|X|$
    or
(ii) when U is one of the order statistics $X_{(1)} \le \cdots \le X_{(n)}$.

Let $X_1, \ldots, X_n$ be a random sample from a probability distribution with p.d.f. $f_X(x)$ and c.d.f. $F_X(x)$, and let $X_{(1)} \le X_{(2)} \le \ldots \le X_{(n)}$ be the sample variables ordered from least to greatest. We refer to $X_{(1)} \le X_{(2)} \le \ldots \le X_{(n)}$ as the *order statistics* for the random sample. We show how to use the distribution function technique to find the probability distributions for these order statistics.

Example 7.10 Minimum, $X_{(1)}$, or Maximum, $X_{(n)}$, of Random Sample
First, we consider the maximum, $X_{(n)}$, from the sample. Using the fact that the items $X_1, \ldots, X_n$ in the random sample are mutually independent and identically distributed, the c.d.f. for $X_{(n)}$ is given by

$$F_{X_{(n)}}(t) = P\big(X_{(n)} \le t\big) = P(\text{each } X_i \le t) = \prod_{i=1}^{n} P(X_i \le t) = [F_X(t)]^n, \quad -\infty < t < \infty.$$

Note that this result holds whether the underlying probability distribution is continuous or discrete.
    Similarly, for the minimum, $X_{(1)}$, we find

$$F_{X_{(1)}}(t) = P(X_{(1)} \le t) = 1 - P(X_{(1)} > t) = 1 - P(\text{each } X_i > t)$$
$$\stackrel{indep}{=} 1 - \prod_{i=1}^{n} P(X_i > t) = 1 - [1 - F_X(t)]^n, \quad -\infty < t < \infty.$$

Again, this result holds whether the underlying probability distribution is continuous or discrete.

When the underlying distribution is continuous, we can directly obtain the p.d.f.'s for $X_{(n)}$ and $X_{(1)}$ by differentiating the corresponding c.d.f.'s. (We would need to take differences, rather than differentiating, at the corresponding jump points to obtain the p.d.f.'s directly from the c.d.f.'s when the underlying distribution is discrete. This is seldom useful in practice.) The associated p.d.f.'s for $X_{(n)}$ and $X_{(1)}$ when the underlying distribution is continuous are thus given by

$$f_{X_{(n)}}(t) = n[F_X(t)]^{n-1} f_X(t), \quad -\infty < t < \infty,$$

and

$$f_{X_{(1)}}(t) = n[1 - F_X(t)]^{n-1} f_X(t), \quad -\infty < t < \infty.$$

Example 7.11 Continuous Distribution
Let $X_1, \ldots, X_n$ be a random sample from the continuous probability distribution with p.d.f.:

$$f_X(x) = 5x^4 I_{(0,1)}(x)$$

and c.d.f.

$$F_X(x) = 0, \quad x \le 0,$$
$$= x^5, \quad 0 < x < 1,$$
$$= 1, \quad x \ge 1.$$

Then, the c.d.f. and p.d.f. for the maximum, $X_{(n)}$, are given by

$$F_{X_{(n)}}(x) = 0, \quad x \le 0,$$
$$= x^{5n}, \quad 0 < x < 1,$$
$$= 1, \quad x \ge 1$$

and

$$f_{X_{(n)}}(x) = 5nx^{5n-1}I_{(0,1)}(x),$$

respectively. It follows that

$$E(X_{(n)}) = \int_0^1 x5nx^{5n-1}\,dx = \frac{5nx^{5n+1}}{5n+1}\Big|_{x=0}^{x=1} = \frac{5n}{5n+1}.$$

Example 7.12 Logistic Distribution
Let $X_1,\ldots,X_n$ be a random sample from the continuous probability distribution with p.d.f.

$$f_X(x) = \frac{e^{-x}}{[1+e^{-x}]^2}, \quad -\infty < x < \infty.$$

(This is called the *logistic distribution*.) The associated c.d.f. is

$$F_X(x) = (1+e^{-x})^{-1}, \quad -\infty < x < \infty.$$

It follows that the c.d.f. for the minimum, $X_{(1)}$, is given by

$$F_{X_{(1)}}(x) = 1 - \left[1 - (1+e^{-x})^{-1}\right]^n = 1 - \left[\frac{e^{-x}}{1+e^{-x}}\right]^n$$

$$= 1 - \left[\frac{1}{1+e^x}\right]^n = 1 - \left(1+e^x\right)^{-n}, \quad -\infty < x < \infty,$$

and the corresponding p.d.f. for $X_{(1)}$ is

$$f_{X_{(1)}}(x) = \frac{d}{dx}F_{X_{(1)}}(x) = \frac{d}{dx}\{1 - (1+e^x)^{-n}\}$$

$$= n(1+e^x)^{-n-1}e^x = \frac{ne^x}{(1+e^x)^{n+1}}, \quad -\infty < x < \infty.$$

Example 7.13 Arbitrary Order Statistic
Now we proceed to finding expressions for the c.d.f. and p.d.f. for an arbitrary order statistic, $X_{(j)}$, from a random sample of size n. Let $j \in \{1,\ldots,n\}$ be arbitrary. Then the c.d.f. for $X_{(j)}$ is given by

$$F_{X_{(j)}}(x) = P(X_{(j)} \le x) = P(\text{at least } j\ X\text{'s are } \le x) = \sum_{u=j}^n P(\text{exactly } u\ X\text{'s are } \le x)$$

$$\underset{\text{Binomial } n,\, p=P(X\le x)=F_X(x)}{=} \sum_{u=j}^n \binom{n}{u} [F_X(x)]^u [1 - F_X(x)]^{n-u}.$$

Thus, for any $j \in \{1,\ldots,n\}$, the c.d.f. for $X_{(j)}$ is

$$F_{X_{(j)}}(x) = \sum_{u=j}^{n} \binom{n}{u} [F_X(x)]^u [1 - F_X(x)]^{n-u}, \quad -\infty < x < \infty.$$

This expression holds for both continuous and discrete underlying distributions.

In the case of an underlying continuous distribution, we can once again obtain the p.d.f. for $X_{(j)}$ by differentiating this expression for the c.d.f. Doing so, we have

$$f_{X_{(j)}}(x) = \frac{d}{dx} F_{X_{(j)}}(x) = \frac{d}{dx} \sum_{u=j}^{n} \binom{n}{u} [F_X(x)]^u [1 - F_X(x)]^{n-u}$$

$$= \sum_{u=j}^{n} \binom{n}{u} \frac{d}{dx} \{ [F_X(x)]^u [1 - F_X(x)]^{n-u} \}$$

$$= \sum_{u=j}^{n} \binom{n}{u} \left\{ u[F_X(x)]^{u-1} f_X(x)[1 - F_X(x)]^{n-u} + [F_X(x)]^u (n-u)[1 - F_X(x)]^{n-u-1}(-f_X(x)) \right\}$$

$$= \sum_{u=j}^{n} \frac{n!}{(u-1)!(n-u)!} [F_X(x)]^{u-1} [1 - F_X(x)]^{n-u} f_X(x)$$

$$- \sum_{u=j}^{n-1} \frac{n!}{u!(n-u-1)!} [F_X(x)]^u [1 - F_X(x)]^{n-u-1} f_X(x)$$

$$= A - B,$$

(7.1)

where we note that the $u = n$ term in B is not included since $(n - u) = 0$ for that term.

Now, we expand A to obtain

$$A = \frac{n!}{(j-1)!(n-j)!} [F_X(x)]^{j-1} [1 - F_X(x)]^{n-j} f_X(x)$$

$$+ \sum_{u=j+1}^{n} \frac{n!}{(u-1)!(n-u)!} [F_X(x)]^{u-1} [1 - F_X(x)]^{n-u} f_X(x)$$

$$= \frac{n!}{(j-1)!(n-j)!} [F_X(x)]^{j-1} [1 - F_X(x)]^{n-j} f_X(x) + C.$$

But, letting $v = u - 1$ in the summation for C, we see that

$$C = \sum_{u=j+1}^{n} \frac{n!}{(u-1)!(n-u)!} [F_X(x)]^{u-1} [1 - F_X(x)]^{n-u} f_X(x)$$

$$= \sum_{v=j}^{n-1} \frac{n!}{v!(n-v-1)!} [F_X(x)]^v [1 - F_X(x)]^{n-v-1} f_X(x) = B,$$

canceling with the negative B in (7.1) so that the final expression for the p.d.f. of $X_{(j)}$ is given by

$$f_{X_{(j)}}(x) = \frac{n!}{(j-1)!(n-j)!} [F_X(x)]^{j-1} [1 - F_X(x)]^{n-j} f_X(x), \quad -\infty < x < \infty.$$

Note the very intuitive nature of this expression for the p.d.f. of the jth order statistic in a random sample of size n from $f_X(x)$.

Example 7.14 Uniform Distribution
Let $X_1,\dots,X_n$ be a random sample from the $Unif(0, 1)$ distribution with p.d.f.

$$f_X(x) = 1\, I_{(0,1)}(x)$$

and c.d.f.

$$F_X(x) = 0, \quad x \le 0$$

$$= \int_0^x dt = x, \quad 0 < x < 1$$

$$= 1, \quad x \ge 1.$$

Thus, for $X \sim Unif(0, 1)$, the p.d.f. for the jth order statistic, $X_{(j)}$, is given by

$$f_{X_{(j)}}(x) = \frac{n!}{(j-1)!(n-j)!} x^{j-1}(1-x)^{n-j} I_{(0,1)}(x),$$

or, using gamma function notation,

$$f_{X_{(j)}}(x) = \frac{\Gamma(n+1)}{\Gamma(j)\Gamma(n-j+1)} x^{j-i}(1-x)^{n-j} I_{(0,1)}(x).$$

Think About It   Note that in this $Unif(0, 1)$ setting, $X_{(j)}$ and $X_{(n-j+1)}$ have the same distributional form for all $j = 1, \dots, n$. Does this make intuitive sense in this setting?

The distribution for $X_{(j)}$ in Example 7.14 is a special case of a class of distributions, called the *Beta distributions*, which we now define.

Definition 7.5   A random variable X is said to have a **Beta distribution with parameters $\alpha > 0$ and $\beta > 0$** if it has p.d.f.

$$f_X(x) = \frac{\Gamma(\alpha + \beta)}{\Gamma(\alpha)\Gamma(\beta)} x^{\alpha-1}(1-x)^{\beta-1} I_{(0,1)}(x) \qquad (7.2)$$

We denote this by $X \sim Beta(\alpha, \beta)$.

Note The expression for the Beta p.d.f. in (7.2) implies that

$$\int_0^1 x^{\alpha-1}(1-x)^{\beta-1}dx = \frac{\Gamma(\alpha)\Gamma(\beta)}{\Gamma(\alpha+\beta)},$$

for any $\alpha > 0$ and $\beta > 0$, a fact that we will find useful throughout the rest of the text.

From Definition 7.5 and Example 7.14, we see that the jth order statistic, $X_{(j)}$, for a random sample of size n from the *Unif* (0, 1) distribution has a *Beta* $(j, n-j+1)$ distribution.

Example 7.15 Properties of a Beta Distribution
Let $X \sim Beta\ (\alpha, \beta)$. Then

$$E[X] = \int_0^1 x \frac{\Gamma(\alpha+\beta)}{\Gamma(\alpha)\Gamma(\beta)} x^{\alpha-1}(1-x)^{\beta-1}dx = \frac{\Gamma(\alpha+\beta)}{\Gamma(\alpha)\Gamma(\beta)} \int_0^1 x^{(\alpha+1)-1}(1-x)^{\beta-1}dx$$

$$= \frac{\Gamma(\alpha+\beta)}{\Gamma(\alpha)} \frac{\Gamma(\alpha+1)}{\Gamma(\alpha+\beta+1)} \int_0^1 \frac{\Gamma(\alpha+\beta+1)}{\Gamma(\alpha+1)\Gamma(\beta)} x^{(\alpha+1)-1}(1-x)^{\beta-1}dx \qquad (7.3)$$

$$= \frac{\Gamma(\alpha+\beta)}{\Gamma(\alpha)} \frac{\Gamma(\alpha+1)}{\Gamma(\alpha+\beta+1)} = \frac{\alpha}{\alpha+\beta}$$

since the final integral in (7.3) is just the integral of the *Beta* $(\alpha + 1, \beta)$ p.d.f. You are asked to show in Exercise 7.19 that

$$Var(X) = \frac{\alpha\beta}{(\alpha+\beta)^2(\alpha+\beta+1)}.$$

Returning to Example 7.14, we now see that for the *Unif* (0, 1) distribution,

$$E[X_{(j)}] = \frac{j}{j+(n-j+1)} = \frac{j}{n+1}$$

and

$$Var(X_{(j)}) = \frac{j(n-j+1)}{(n+1)^2(n+2)}.$$

Think About It Do these expressions make intuitive sense for the *Unif* (0, 1) distribution?

7.3.3 *Change of Variable Technique*

Let U_1 denote the statistic of interest $(U(X_1, \ldots, X_n))$ and define $n - 1$ additional random variables $U_2 = U_2(X_1, \ldots, X_n), \ldots, U_n = U_n(X_1, \ldots, X_n)$. (While you have complete freedom in how to choose these $n - 1$ additional variables, as we shall see, this should be done carefully to make your task easier!) In this text, we consider only the case where these n random variables represent a 1-1 transformation from the sample space for the X's to the sample space for the U's. In that setting, let

$$x_1 = w_1(u_1, \ldots, u_n), \cdots, x_n = w_n(u_1, \ldots, u_n)$$

be the unique inverse functions.

Setting 1: Change of Variable for Discrete Variables
For an underlying discrete distribution, the joint p.d.f. for $X_1, \ldots, X_n$ is given by

$$f_{X_1, \ldots, X_n}(x_1, \ldots, x_n) = \prod_{i=1}^{n} f_X(x_i)$$

and the associated joint p.d.f. for $U_1, \ldots, U_n$ is

$$g_{U_1, \ldots, U_n}(u_1, \ldots, u_n) = \prod_{i=1}^{n} f_X(w_i(u_1, \ldots, u_n)).$$

It follows that the marginal p.d.f. for the statistic of interest $U = U_1$ is obtained by summing out the other $n - 1$ variables to obtain

$$g_U(u_1) = \sum_{(u_2, \ldots, u_n)} \sum \cdots \sum g_{U_1, \ldots, U_n}(u_1, \ldots, u_n).$$

Setting 2: Change of Variable for Continuous Variables
Let J represent the Jacobian determinant

$$J = \begin{vmatrix} \dfrac{\partial x_1}{\partial u_1} & \cdots & \dfrac{\partial x_1}{\partial u_n} \\ & \vdots & \\ \dfrac{\partial x_n}{\partial u_1} & \cdots & \dfrac{\partial x_n}{\partial u_n} \end{vmatrix} \neq 0.$$

Then the joint p.d.f. for $U_1, \ldots, U_n$ in this continuous setting is

$$g_{U_1, \ldots, U_n}(u_1, \ldots, u_n) = \prod_{i=1}^{n} f_X(w_i(u_1, \ldots, u_n)) \, | \, J \, |,$$

and the marginal p.d.f for the statistic of interest $U = U_1$ is obtained by integrating
out the other $n - 1$ variables to obtain

$$g_U(u_1) = \int_{-\infty}^{\infty} \int_{-\infty}^{\infty} \cdots \int_{-\infty}^{\infty} g_{U_1,\ldots,U_n}(u_1, \ldots, u_n)du_2 \ldots u_n.$$

Note The change of variable approach will always work, but the necessary mathematical details can be extensive and involved. In this book, we illustrate this change
of variable approach only for the case where $n = 2$.

Example 7.16 Discrete Setting, Two Poisson Variables
Let X_1 and X_2 be independent random variables having Poisson distributions with
parameters λ_1 and λ_2, respectively. Then the joint p.d.f. for X_1 and X_2 is

$$f_{X_1,X_2}(x_1,x_2) = \frac{\lambda_1^{x_1} e^{-\lambda_1}}{x_1!} \frac{\lambda_2^{x_2} e^{-\lambda_2}}{x_2!} = \frac{\lambda_1^{x_1} \lambda_2^{x_2} e^{-\lambda_1-\lambda_2}}{x_1!x_2!} \prod_{i=1}^{2} I_{\{0,1,2,\ldots\}}(x_i).$$

We want to obtain the p.d.f. for the variable $U = U_1 = X_1 + X_2$ using the change of
variable technique. In order to do this, we must first define a second variable, U_2.
While this second variable is completely arbitrary, the entire process can be simpli-
fied by making a strategic choice for U_2. In particular, we opt here to take $U_2 = X_2$.
(Feel free to try some other choices!) Thus, the pair $(U_1, U_2) = (X_1 + X_2, X_2)$
represents a one-to-one transformation from (X_1, X_2) with unique inverse functions

$$X_1 = U_1 - U_2 \text{ and } X_2 = U_2$$

that maps the joint space $A = \prod_{i=1}^{2} I_{\{0,1,2,\ldots\}}(x_i)$ for (X_1, X_2) onto the joint space $B =$
$I_{\{0,1,2,\ldots\}}(u_1)I_{\{0,1,\ldots,u_1\}}(u_2)$ for (U_1, U_2). Then the joint p.d.f. for (U_1, U_2) is given by

$$g_{U_1,U_2}(u_1,u_2) = f_{X_1,X_2}(u_1 - u_2, u_2) = \frac{\lambda_1^{u_1 - u_2} \lambda_2^{u_2} e^{-\lambda_1-\lambda_2}}{(u_1 - u_2)!u_2!} I_B(u_1, u_2).$$

The marginal p.d.f. for $U = U_1 = X_1 + X_2$ is obtained then by summing this
expression over the possible values for U_2; that is, the marginal p.d.f. for $U = U_1$ is

$$g_U(u) = \sum_{u_2=0}^{u} g_{U_1,U_2}(u,u_2) = \sum_{u_2=0}^{u} \frac{\lambda_1^{u-u_2} \lambda_2^{u_2} e^{-\lambda_1-\lambda_2}}{(u - u_2)!u_2!}$$

$$= \frac{e^{-\lambda_1-\lambda_2}}{u!} \sum_{u_2=0}^{u} \frac{u!}{(u - u_2)!u_2!} \lambda_1^{u-u_2} \lambda_2^{u_2}$$

$$= \frac{e^{-\lambda_1-\lambda_2}}{u!} (\lambda_1 + \lambda_2)^u \sum_{u_2=0}^{u} \frac{u!}{u_2!(u - u_2)!} \left(\frac{\lambda_1}{\lambda_1 + \lambda_2}\right)^{u-u_2} \left(\frac{\lambda_2}{\lambda_1 + \lambda_2}\right)^{u_2},$$

where, to get this last expression, we have simply multiplied the right-hand side of the previous expression by the "fancy" $1 = \frac{(\lambda_1+\lambda_2)^u}{(\lambda_1+\lambda_2)^u}$, with the numerator kept outside the summation and the denominator taken inside the summation. It follows that

$$g_U(u) = \frac{e^{-(\lambda_1+\lambda_2)}}{u!} (\lambda_1 + \lambda_2)^u \sum_{u_2=0}^{u} \binom{u}{u_2} \left(\frac{\lambda_2}{\lambda_1 + \lambda_2}\right)^{u_2} \left(1 - \frac{\lambda_2}{\lambda_1 + \lambda_2}\right)^{u-u_2}.$$

Since $0 \le \frac{\lambda_2}{\lambda_1+\lambda_2} \le 1$, the sum in this expression is just the summation over the entire sample space for a *Binom* $\left(u, \frac{\lambda_2}{\lambda_2+\lambda_1}\right)$ random variable and, hence, is nothing more than another useful representation of 1. (Keep this idea of alternative, more complex, expressions for 1 in your toolbox as you approach solving other statistical or mathematical problems.) It follows that the marginal p.d.f. for U is

$$g_U(u) = \frac{(\lambda_1 + \lambda_2)^u \, e^{-(\lambda_1+\lambda_2)}}{u!} \, I_{\{0,1,2,\ldots\}}(u),$$

which we recognize to be the p.d.f. for a Poisson random variable with parameter $\lambda_1 + \lambda_2$. Thus, we have established the fact that the sum of two independent Poisson random variables with parameters λ_1 and λ_2, respectively, is also Poisson distributed with parameter $\lambda_1 + \lambda_2$.

Think About It Do you think that this result extends to the sum of an arbitrary number, say n, of independent Poisson random variables? The answer is yes, and we will return to our old friend induction to prove that result.

Theorem 7.1 Sum of Independent Poisson Variables
Let $X_1,\ldots,X_n$ be mutually independent random variables such that $X_i \sim$ *Poisson* (λ_i), for $i = 1, \ldots, n$. Then, $V = \sum_{i=1}^{n} X_i \sim$ *Poisson* $(\lambda_1 + \cdots + \lambda_n)$.

Proof As promised, we use induction to establish this result. Let A_k, $k = 1, 2, \ldots$, be the event defined by

$$A_k : X_1 + (X_2 + \cdots + X_{k+1}) \sim Poisson \, (\lambda_1 + \cdots + \lambda_{k+1}).$$

Consider $k = 1$. A_1 corresponds to the event that the sum of two independent Poisson variables with parameters λ_1 and λ_2 is also Poisson with parameter $\lambda_1 + \lambda_2$, which is the result that we obtained in Example 7.16.

Consider $k = n - 1$ and assume that $A_k = A_{n-1}$ is true, that is,

$$X_1 + (X_2 + \cdots + X_n) = X_1 + \cdots + X_n \sim Poisson \, (\lambda_1 + \cdots + \lambda_n)$$

Consider $k = n$. The random variable associated with the event A_n is

$$X_1 + \cdots + X_n + X_{n+1} = Y + X_{n+1},$$

where $Y = X_1 + \cdots + X_n$. From the inductive assumption for $k = n - 1$, we know that $Y \sim Poisson\ (\lambda_1 + \cdots + \lambda_n)$. Moreover, Y and X_{n+1} are independent. Using Example 7.16 once again, we have

$$Y + X_{n+1} = X_1 + \cdots + X_n + X_{n+1} \sim Poisson\ (\lambda_1 + \cdots + \lambda_n + \lambda_{n+1}),$$

which establishes the fact that the event A_n is also true. Therefore, it follows from induction that the sum of an arbitrary number of independent Poisson variables is also a Poisson variable with parameter equal to the sum of the individual Poisson parameters. (Compare this with Exercise 7.13, where you are asked to show the same result using the moment generating function approach. Which method of proof do you prefer for this problem?) ∎

Think About It Do you think that the Poisson distribution played a special role in the induction approach to proving Theorem 7.1? Suppose that $X_1,\ldots,X_n$ are independent random variables with the same distributional form but possibly with different parameters $\theta_1,\ldots,\theta_n$. If we can show that $X_1 + X_2$ has the same distributional form with parameter $\theta_1 + \theta_2$, does that tell us anything about the distribution of $X_1 + \cdots + X_n$? Discuss how this general induction result can be applied to the setting where the underlying distributional form is binomial with common p value and possibly different numbers of trials $m_1,\ldots,m_n$ or to the setting where the underlying distributional form is normal with common variance σ^2 and possibly different means $\mu_1,\ldots,\mu_n$.

Example 7.17 Continuous Setting, Two Gamma Variables
Let X_1 and X_2 be independent random variables with $X_i \sim Gamma\ (\alpha_i, \beta)$, for $i = 1$, 2, where $\alpha_1 > 0$, $\alpha_2 > 0$, *and* $\beta > 0$. Thus, the joint p.d.f. for (X_1, X_2) is given by

$$f_{X_1,X_2}(x_1, x_2) = \frac{x_1^{\alpha_1-1} x_2^{\alpha_2-1}}{\Gamma(\alpha_1)\Gamma(\alpha_2)\beta^{\alpha_1+\alpha_2}} e^{-\left(\frac{x_1+x_2}{\beta}\right)} I_{(0,\infty)}(x_1) I_{(0,\infty)}(x_2).$$

Define the new variables $Y_1 = X_1 + X_2$ and $Y_2 = \frac{X_1}{X_1+X_2}$. This represents a 1-1 transformation from the (X_1, X_2) space

$$A = \{(x_1, x_2) : 0 < x_1 < \infty, 0 < x_2 < \infty\}$$

onto the (Y_1, Y_2) space

$$B = \{(y_1, y_2) : 0 < y_1 < \infty, 0 < y_2 < 1\},$$

with unique inverse functions $x_1 = y_1 y_2$ and $x_2 = y_1 - x_1 = y_1(1 - y_2)$. The associated Jacobian is given by

$$\text{Jacobian} = J = \det \begin{bmatrix} \dfrac{\partial x_1}{\partial y_1} & \dfrac{\partial x_1}{\partial y_2} \\[2mm] \dfrac{\partial x_2}{\partial y_1} & \dfrac{\partial x_2}{\partial y_2} \end{bmatrix} = \det \begin{bmatrix} y_2 & y_1 \\ (1-y_2) & -y_1 \end{bmatrix} = -y_2 y_1 - y_1 + y_2 y_1$$

$$= -y_1.$$

$\Rightarrow |J| = |-y_1| = y_1$, and the joint p.d.f. for (Y_1, Y_2) is given by

$$g_{(Y_1,Y_2)}(y_1,y_2) = f_{X_1,X_2}(y_1 y_2, y_1(1-y_2)) \, |-y_1|$$

$$= \frac{(y_1 y_2)^{\alpha_1-1}\{y_1(1-y_2)\}^{\alpha_2-1}}{\Gamma(\alpha_1)\Gamma(\alpha_2)\beta^{\alpha_1}\beta^{\alpha_2}} e^{-\frac{\{y_1 y_2 + y_1(1-y_2)\}}{\beta}} \, |-y_1| \, I_{(0,\infty)}(y_1)\, I_{(0,1)}(y_2)$$

$$= \frac{y_1^{\alpha_1+\alpha_2-1} y_2^{\alpha_1-1}(1-y_2)^{\alpha_2-1}}{\Gamma(\alpha_1)\Gamma(\alpha_2)\beta^{\alpha_1+\alpha_2}} e^{-\frac{y_1}{\beta}} I_{(0,\infty)}(y_1) I_{(0,1)}(y_2)$$

$$= \left\{ \frac{y_1^{\alpha_1+\alpha_2-1} e^{-\frac{y_1}{\beta}}}{\Gamma(\alpha_1+\alpha_2)\beta^{\alpha_1+\alpha_2}} I_{(0,\infty)}(y_1) \right\} \left\{ \frac{\Gamma(\alpha_1+\alpha_2)}{\Gamma(\alpha_1)\Gamma(\alpha_2)} y_2^{\alpha_1-1}(1-y_2)^{\alpha_2-1} I_{(0,1)}(y_2) \right\}$$

$$= g_{Y_1}(y_1) g_{Y_2}(y_2).$$

(Notice how we were effectively able to use the "funny" 1, $\frac{\Gamma(\alpha_1+\alpha_2)}{\Gamma(\alpha_1+\alpha_2)}$.) Thus, it follows (surprisingly!) that Y_1 and Y_2 are independent random variables with p.d.f.'s

$$g_{Y_1}(y_1) = \frac{y_1^{\alpha_1+\alpha_2-1} e^{-\frac{y_1}{\beta}}}{\Gamma(\alpha_1+\alpha_2)\beta^{\alpha_1+\alpha_2}} I_{(0,\infty)}(y_1)$$

and

$$g_{Y_2}(y_2) = \frac{\Gamma(\alpha_1+\alpha_2)}{\Gamma(\alpha_1)\Gamma(\alpha_2)} y_2^{\alpha_1-1}(1-y_2)^{\alpha_2-1} I_{(0,1)}(y_2),$$

respectively. Since we recognize both of these p.d.f.'s, we can conclude that

$$Y_1 \sim Gamma\ (\alpha_1 + \alpha_2,\ \beta),$$
$$Y_2 \sim Beta\ (\alpha_1, \alpha_2),$$

and

$$Y_1 \text{ and } Y_2 \text{ are independent.}$$

Note that Example 7.17 implies the following:

1. The sum of independent gamma variables with parameters (α_1, β) and (α_2, β) also has a gamma distribution with parameters $(\alpha_1 + \alpha_2, \beta)$.

2. The ratio of a gamma variable with parameters (α_1, β) divided by the sum of itself plus a second independent gamma variable with parameters (α_2, β) has a Beta distribution with parameters (α_1, α_2).
3. Surprisingly, the ratio is independent of its own denominator!

Think About It These results place no restrictions on either α_1 or α_2, but they do assume that the independent gamma variables have a common second parameter β. Do you think this is necessary?

Think About It Do you think that a result similar to Example 7.17 holds for more than two independent gamma variables with common second parameter β? This question is explored in the exercises for this chapter.

Remark In Example 7.17, we found that $Y_1 = X_1 + X_2$ and $Y_2 = \frac{X_1}{X_1 + X_2}$ were independent random variables. From this, it follows that

$$E[Y_1 Y_2] = E\left[(X_1 + X_2)\left\{\frac{X_1}{X_1 + X_2}\right\}\right] = E[X_1] \overset{ind}{=} E[Y_1]E[Y_2]$$

$$= E[X_1 + X_2]E\left[\frac{X_1}{X_1 + X_2}\right],$$

which implies the unusual fact that

$$E\left[\frac{X_1}{X_1 + X_2}\right] = \frac{E[X_1]}{E[X_1 + X_2]}.$$

Thus, using properties of gamma distributions, it follows that

$$E\left[\frac{X_1}{X_1 + X_2}\right] = \frac{\alpha_1 \beta}{(\alpha_1 + \alpha_2)\beta} = \frac{\alpha_1}{\alpha_1 + \alpha_2},$$

which is not a surprise, since $Y_2 = \frac{X_1}{X_1 + X_2} \sim Beta(\alpha_1, \alpha_2)$. (See Example 7.15.)

Caution! We need to emphasize that this unusual relationship (namely, $E\left[\frac{X_1}{X_1 + X_2}\right] = \frac{E[X_1]}{E[X_1 + X_2]}$) is valid ONLY because the ratio $\frac{X_1}{X_1 + X_2}$ is independent of its denominator $X_1 + X_2$. Do *not* try to make your life simpler by applying this to an arbitrary ratio of two random variables without this special independence feature.

7.4 Equal in Distribution Approach to Obtaining Properties of Sampling Distributions

In some settings, we are not necessarily interested in the full sampling distribution for a relevant statistic. At times, it suffices simply to know certain properties of the sampling distribution for the statistic, such as its mean or variance, or even whether the sampling distribution is symmetric. For such situations, an interesting approach, called the *equal in distribution technique*, can be quite helpful.

Definition 7.6 Let S and T be random variables with c.d.f.'s $F(s)$ and $G(t)$, respectively. We say that S and T are **equal in distribution**, and write $S \overset{d}{=} T$, if $F(x) \equiv G(x)$.

Extension Let $X = (X_1, \ldots, X_n)$ and $Y = (Y_1, \ldots, Y_n)$ be n-dimensional random vectors. We say that X and Y are equal in distribution, and write $X \overset{d}{=} Y$, if X and Y have the same joint n-dimensional distribution.

We first establish an equal in distribution result about a symmetric distribution.

Theorem 7.2
A random variable X has a probability distribution that is symmetric about the point μ if and only if $X - \mu \overset{d}{=} \mu - X$.

Proof Let $F(x)$ denote the c.d.f. for X. First, assume that $X - \mu \overset{d}{=} \mu - X$. Then

$$F(\mu + t) = P(X \leq \mu + t) = P(X - \mu \leq t) = P(\mu - X \leq t)$$
$$= P(X \geq \mu - t) = 1 - F((\mu - t)^-) \quad \text{for all } t$$

$\Rightarrow F(\mu + t) + F((\mu - t)^-) = 1$ for all $t \Rightarrow$ the X distribution is symmetric about μ. Now, assume that the X distribution is symmetric about μ. Then, we have

$$F(\mu + t) + F((\mu - t)^-) = 1 \text{ for all } t.$$

The c.d.f. for $X - \mu$ is given by

$$G(t) = P(X - \mu \leq t) = P(X \leq \mu + t) = F(\mu + t) \text{ for all } t,$$

and the c.d.f. for $\mu - X$ is

$$H(t) = P(\mu - X \leq t) = P(X \geq \mu - t) = 1 - F((\mu - t)^-) \quad \text{for all } t.$$

By our symmetry assumption, it follows that

$$H(t) = G(t) \text{ for all } t,$$

which is equivalent to $X - \mu \overset{d}{=} \mu - X$, and the proof is complete. ∎

Corollary 7.1 Let X have a symmetric distribution about the point μ. Then, $E[X] = \mu$, if it exists.

Proof From Theorem 7.2, the symmetry assumption implies that $X - \mu \overset{d}{=} \mu - X$. Taking expectations (assumed to exist) of both sides of this equal in distribution expression yields

$$E[X] - \mu = \mu - E[X] \Rightarrow E[X] = \mu. \ \blacksquare$$

We now state (without proof) an important and powerful result for equal in distribution arguments. While the result should make intuitive sense, a formal proof requires some knowledge of measure theory.

Theorem 7.3

Let $\underset{\sim}{X} = (X_1, \ldots, X_n)$ and $\underset{\sim}{Y} = (Y_1, \ldots, Y_n)$ be n-dimensional random vectors such that $\underset{\sim}{X} \overset{d}{=} \underset{\sim}{Y}$. Let $U(\cdot)$ be a "nice" (measurable) function (possibly vector valued as well) defined on the common support of $\underset{\sim}{X}$ and $\underset{\sim}{Y}$. Then $U\left(\underset{\sim}{X}\right) \overset{d}{=} U\left(\underset{\sim}{Y}\right)$.

Example 7.18 Independent, Identically Distributed Continuous Variables

Let X and Y be independent, identically distributed continuous random variables. Then $(X, Y) \overset{d}{=} (Y, X)$. Define the function $\delta(x, y) = 1, 0$ if $y >, \leq x$. Then, by Theorem 7.3, we know that $\delta(X, Y) \overset{d}{=} \delta(Y, X)$. It follows that

$$E[\delta(X, Y)] = P(Y - X > 0) = P(Y > X) = E[\delta(Y, X)] = P(X - Y > 0)$$
$$= P(X > Y).$$

Since $Y - X$ is a continuous random variable (so that $P(Y = X) = 0$), we have

$$P(X < Y) = P(X > Y) = 1/2.$$

While this is not a surprising result, its proof is immediate through this equal in distribution argument.

Example 7.19 Symmetrically Situated Order Statistics

Let $X_{(1)} \leq X_{(2)} \leq \cdots \leq X_{(n)}$ denote the order statistics for a random sample $X_1, \ldots, X_n$ from a continuous distribution that is symmetric about the point μ. From Exercise 7.21, we know that

$$(X_1 - \mu, \ldots, X_n - \mu) \overset{d}{=} (\mu - X_1, \ldots, \mu - X_n). \tag{7.4}$$

Define the function $U(t_1, \ldots, t_n)$ on the n-dimensional real numbers by

$$U(t_1, \ldots, t_n) = \left(t_{(1)}, \ldots, t_{(n)}\right),$$

where $t_{(1)} \leq t_{(2)} \leq \cdots \leq t_{(n)}$ are the ordered values of $t_1, \ldots, t_n$. Applying this $U(\cdot)$ function to both sides of the equal in distribution statement in (7.4), we see from Theorem 7.3 that

$$\left((X - \mu)_{(1)}, (X - \mu)_{(2)}, \ldots, (X - \mu)_{(n)}\right) \overset{d}{=} \left((\mu - X)_{(1)}, (\mu - X)_{(2)}, \ldots, (\mu - X)_{(n)}\right).$$

However,

$$(X - \mu)_{(j)} = X_{(j)} - \mu, \text{ for } j = 1, \ldots, n$$

and

$$(\mu - X)_{(j)} = \mu - X_{(n+1-j)}, \text{ for } j = 1, \ldots, n.$$

Thus, we have

$$\left(X_{(1)} - \mu, X_{(2)} - \mu, \ldots, X_{(n)} - \mu\right) \overset{d}{=} \left(\mu - X_{(n)}, \mu - X_{(n-1)}, \ldots, \mu - X_{(1)}\right), \quad (7.5)$$

providing a nice relationship between symmetrically situated order statistics from an underlying continuous, symmetric distribution.

Now, apply a second function $V(t_1, \ldots, t_n) = t_j$, for $j \in \{1, 2, \ldots, n\}$, to the new equal in distribution relationship in (7.5). This yields the following marginal relationship between symmetrically situated order statistics from an underlying continuous, symmetric distribution:

$$\left(X_{(j)} - \mu\right) \overset{d}{=} \left(\mu - X_{(n-j+1)}\right), \text{ for } j = 1, \ldots, n. \quad (7.6)$$

Taking expectations (provided they exist) on both sides of this equal in distribution expression, we obtain

$$E\left[X_{(j)} - \mu\right] = E\left[\mu - X_{(n-j+1)}\right], \text{ for } j = 1, \ldots, n,$$

which, in turn, implies that

$$E\left[X_{(j)}\right] = 2\mu - E\left[X_{(n-j+1)}\right]$$

and

$$E\left[\frac{X_{(j)} + X_{(n+1-j)}}{2}\right] = \mu,$$

for $j = 1, \ldots, n$.

Thus, each of the symmetrically situated averages $\frac{X_{(j)} + X_{(n+1-j)}}{2}$, $j = 1, \ldots, n$, is an unbiased estimator for the point of symmetry, μ.

Example 7.20 Sample Mean from a Symmetric Distribution

Let $X_1,\ldots,X_n$ be a random sample from a distribution that is symmetric about the point μ. Applying the function $U(t_1, \ldots, t_n) = \sum_{i=1}^{n} t_i/n$ to both sides of the equal in distribution relationship in (7.4), we see from Theorem 7.3 that

$$\sum_{i=1}^{n}(X_i - \mu)/n \overset{d}{=} \sum_{j=1}^{n}(\mu - X_j)/n \Rightarrow (\overline{X} - \mu) \overset{d}{=} (\mu - \overline{X}), \qquad (7.7)$$

where $\overline{X} = \sum_{i=1}^{n} X_i/n$ is the sample mean. It follows from this equal in distribution relationship in (7.7) that the distribution of $\overline{X}$ is also symmetric about μ. Moreover, if the expectation exists, this implies that $E[\overline{X}] = \mu$, which we already knew even if the underlying distribution is not symmetric.

Examples 7.19 and 7.20 (as well as Exercises 7.22 and 7.23) suggest that there might be a more general result for showing that a statistic is unbiased for the point of symmetry of a distribution. Indeed, there is, and we now introduce the necessary notation and discussion to establish this general result.

Definition 7.7 Let $t(X_1, \ldots, X_n)$ be a statistic based on the sample data $X_1, \ldots, X_n$. The statistic $t(\cdot)$ is said to be:

(i)   an **odd statistic** if

$$t(-x_1, \ldots, -x_n) = -t(x_1, \ldots, x_n)$$

or

(ii)   an **even statistic** if

$$t(-x_1, \ldots, -x_n) = t(x_1, \ldots, x_n)$$

for every $x_1, \ldots, x_n$.

Definition 7.8 Let $t(X_1, \ldots, X_n)$ be a statistic based on the sample data $X_1, \ldots, X_n$. The statistic $t(\cdot)$ is said to be:

(i)   a **translation statistic** if

$$t(x_1 + k, \ldots, x_n + k) = t(x_1, \ldots, x_n) + k$$

or

(continued)

Definition 7.8 (continued)

(ii) a **translation-invariant statistic** if

$$t(x_1 + k, \ldots, x_n + k) = t(x_1, \ldots, x_n)$$

for every k and $x_1, \ldots, x_n$.

Example 7.21 Sample Mean

The sample mean $\overline{X} = \frac{1}{n} \sum_{i=1}^{n} X_i$ is an odd translation statistic, since

$$\frac{1}{n} \sum_{i=1}^{n} (-x_i) = -\overline{x}$$

and

$$\frac{1}{n} \sum_{i=1}^{n} (x_i + k) = \overline{x} + k$$

for every k and $x_1, \ldots, x_n$.

Example 7.22 Sample Variance

The sample variance $S^2 = \frac{1}{n-1} \sum_{i=1}^{n} \left(X_i - \overline{X}\right)^2$ is an even translation-invariant statistic, since

$$\frac{1}{n-1} \sum_{i=1}^{n} \left(-x_i - \frac{1}{n} \sum_{j=1}^{n} (-x_j)\right)^2 = s^2$$

and

$$\frac{1}{n-1} \sum_{i=1}^{n} \left(x_i + k - \frac{1}{n} \sum_{j=1}^{n} (x_j + k)\right)^2 = s^2$$

for every k and $x_1, \ldots, x_n$.

We now prove a theorem that provides more general conditions under which a statistic will have a symmetric distribution. Let ϑ^* be the collection of all transformations from the n-dimensional reals onto the n-dimensional reals.

Theorem 7.4

Let $U\left(\underset{\sim}{X}\right) = U(X_1, \ldots, X_n)$ be a real-valued statistic based on the sample $X_1, \ldots,$ X_n. If there exists a transformation $g(\cdot)$ in ϑ^* and a number μ such that

$$U\left(\underset{\sim}{x}\right) - \mu = \mu - U\left(g\left(\underset{\sim}{x}\right)\right)$$

for every $\underset{\sim}{x}$ in the support of $\underset{\sim}{X}$, and

$$g\left(\underset{\sim}{X}\right) \overset{d}{=} \underset{\sim}{X},$$

then $U\left(\underset{\sim}{X}\right)$ is symmetrically distributed about μ.

Proof The condition $g\left(\underset{\sim}{X}\right) \overset{d}{=} \underset{\sim}{X}$ and Theorem 7.3 imply that

$$U\left[g\left(\underset{\sim}{X}\right)\right] \overset{d}{=} U\left(\underset{\sim}{X}\right).$$

Combining this result with the assumption relating $U\left(\underset{\sim}{x}\right)$ and $U\left(g\left(\underset{\sim}{x}\right)\right)$ yields

$$U\left(\underset{\sim}{X}\right) - \mu = \mu - U\left(g\left(\underset{\sim}{X}\right)\right) \overset{d}{=} \mu - U\left(\underset{\sim}{X}\right).$$

It then follows from Theorem 7.2 that $U\left(\underset{\sim}{X}\right)$ is symmetrically distributed about μ. ∎

Corollary 7.2 Let $X_1, \ldots, X_n$ be a random sample from a distribution that is symmetric about μ. Then an odd translation statistic $V(X_1, \ldots, X_n)$ is also symmetrically distributed about μ.

Proof Since the underlying distribution is symmetric about μ, it follows (Exercise 7.21) that

$$(X_1 - \mu, \ldots, X_n - \mu) \overset{d}{=} (\mu - X_1, \ldots, \mu - X_n)$$

and thus that

$$(X_1, \ldots, X_n) \stackrel{d}{=} (2\mu - X_1, \ldots, 2\mu - X_n).$$

If we let $g_1(\cdot)$ map $(x_1, \ldots, x_n)$ into $(2\mu - x_1, \ldots, 2\mu - x_n)$, the odd translation properties imply that

$$V\left(g_1\left(\underset{\sim}{x}\right)\right) = V(2\mu - x_1, \ldots, 2\mu - x_n) = V(-x_1, \ldots, -x_n) + 2\mu$$

$$= -V(x_1, \ldots, x_n) + 2\mu.$$

That is, we have

$$V\left(\underset{\sim}{x}\right) - \mu = \mu - V\left(g_1\left(\underset{\sim}{x}\right)\right)$$

for every $x_1, \ldots, x_n$. By Theorem 7.4, the statistic $V(X_1, \ldots, X_n)$ is symmetrically distributed about μ. ∎

Thus, we see that when sampling from a population that is symmetric about μ, location estimators such as $\overline{X}$ and the sample median $M = \text{median } (X_1, \ldots, X_n)$ are also symmetrically distributed about μ. Thus, if the expectations exist, both $\overline{X}$ and M are unbiased estimators for μ.

We now turn our attention to using equal in distribution arguments to establish useful covariance results for two variables.

Lemma 7.1
If two random variables V and W satisfy

$$(V - \mu, W) \stackrel{d}{=} (\mu - V, W), \tag{7.8}$$

for some constant μ, then, if it exists, $Cov\ (V, W) = 0$.

Proof Taking expectations of the products of both sides of (7.8), we obtain

$$E[(V - \mu)W] = E[(\mu - V)W] \Rightarrow E[VW] - \mu E[W] = \mu E[W] - E[VW] \tag{7.9}$$
$$\Rightarrow E[VW] = \mu E[W].$$

It also follows from (7.8) that the marginal distributions for $V - \mu$ and $\mu - V$ are the same, that is,

$$V - \mu \stackrel{d}{=} \mu - V,$$

which implies (Theorem 7.2) that the distribution of V is symmetric about μ and, hence, that $E[V] = \mu$, since we assumed that the expectations existed. Combining this with (7.9) yields the result

$$E[VW] = E[V]E[W] \Rightarrow Cov(V, W) = 0. \quad \blacksquare$$

Let $\mathbf{X} = (X_1, \ldots, X_n)$ denote a random vector with support set X. Again, let ϑ^* be the collection of all transformations from the n-dimensional reals onto the n-dimensional reals.

Theorem 7.5
Let $V(\mathbf{X})$ and $W(\mathbf{X})$ be two statistics. Suppose there exists a constant μ and some $g(\cdot)$ in ϑ^* such that

$$V(\mathbf{x}) - \mu = \mu - V(g(\mathbf{x})) \text{ for every } \mathbf{x} \in X,$$
$$W(g(\mathbf{x})) = W(\mathbf{x}) \text{ for every } \mathbf{x} \in X,$$

and

$$g(\mathbf{X}) \overset{d}{=} \mathbf{X}.$$

Then

$$[V(\mathbf{X}) - \mu, W(\mathbf{X})] \overset{d}{=} [\mu - V(\mathbf{X}), W(\mathbf{X})].$$

Proof From our assumptions about $g(\cdot)$, we have

$$[V(\mathbf{X}) - \mu, W(\mathbf{X})] = [\mu - V(g(\mathbf{X})), W(g(\mathbf{X}))] \overset{d}{=} [\mu - V(\mathbf{X}), W(\mathbf{X})]. \quad \blacksquare$$

Corollary 7.3 Let $X_1, \ldots, X_n$ be a random sample from a distribution that is symmetric about some point μ. Let $V(X_1, \ldots, X_n)$ be an odd translation statistic and $W(X_1, \ldots, X_n)$ be an even translation-invariant statistic. Then, if it exists,

$$Cov[V(X_1, \ldots, X_n), W(X_1, \ldots, X_n)] = 0.$$

Proof Exercise 7.31.

This corollary shows that, when sampling from a symmetric population, the sample mean $\overline{X}$ and the sample standard deviation S are uncorrelated. This is, of course, a weaker extension to all symmetric distributions of the stronger result that $\overline{X}$ and S are actually *independent* when the underlying population is normal. On the other hand, this corollary also yields the much broader result that *all* odd translation statistics (such as the sample median M, sample mode, midrange $\frac{X_{(1)} + X_{(n)}}{2}$, etc.) and even translation-invariant statistics (such as the sample range $R = X_{(n)} - X_{(1)}$, $V = \frac{1}{n} \sum_{i=1}^{n} |X_i - M|$, etc.) are uncorrelated, if the expectations exist, when sampling from *any* symmetric population.

7.5 Exercises

7.1. Consider the setting for Example 7.3 but suppose we have a random sample of arbitrary size n. What is the likelihood function for this random sample of size n?

7.2. Let $X_1, \ldots, X_n$ be a random sample of size n from a discrete distribution with p.d.f. given by

$$f_X(a_i) = \pi_i \text{ for arbitrary constants } a_1, \ldots, a_k$$
$$= 0, \text{ elsewhere,}$$

where $0 \le \pi_i \le 1$, with $\sum_{i=1}^{k} \pi_i = 1$. What is the likelihood function for this random sample?

7.3. For sampling with replacement (Setting 1) in Example 7.5, find the variance, $\sigma^2_{\tilde{X}}$, for the sampling distribution of the sample median $\tilde{X}$.

7.4. For sampling with replacement (Setting 1) in Example 7.5, find the variance, $\sigma^2_{\overline{X}}$, for the sampling distribution of the sample mean $\overline{X}$. How does this compare to the variance, σ^2_X, for the underlying population? Are you surprised by this result?

7.5. Show that $Var(\overline{X}) = 5.867$ for Setting 2 in Example 7.5 when sampling without replacement between selections.

7.6. Let (X_1, X_2) be random variables with joint p.d.f. $f_{X_1,X_2}(x_1,x_2)$ and suppose that $E[X_1] = \mu_{X_1}$ and $E[X_2] = \mu_{X_2}$ both exist. Show that $E[X_1 + X_2] = E[X_1] + E[X_2] = \mu_{X_1} + \mu_{X_2}$.

7.7. Let (X_1, X_2) be independent random variables with joint p.d.f. $f_{X_1,X_2}(x_1,x_2)$ and marginal p.d.f.'s $f_{X_1}(x_1)$ and $f_{X_2}(x_2)$. Suppose that $Var(X_1) = \sigma^2_{X_1}$ and $Var(X_2) = \sigma^2_{X_2}$ both exist. Show that $Var(X_1 + X_2) = Var(X_1) + Var(X_2) = \sigma^2_{X_1} + \sigma^2_{X_2}$.

7.8. In Exercise 7.7, you were asked to show that the variance of a sum of independent variables is the sum of their individual variances. Do you think this result is true without the assumption that the two variables are independent? Prove the result or construct a counterexample.

7.9. In Exercise 7.6, you were asked to show that the expected value of the sum of two random variables is the sum of their individual expected values (assuming the two expected values exist) regardless of whether or not the two variables are independent. Use induction to extend this result to the arbitrary case of n random variables $(X_1, \ldots, X_n)$, provided all of their individual expected values exist.

7.10. In Exercise 7.7, you were asked to show that the variance of the sum of two independent random variables is the sum of their individual variances, assuming the two variances exist. Use induction to extend this result to the arbitrary case of n independent random variables $(X_1, \ldots, X_n)$, provided all of their individual variances exist.

7.11. Let $X_1, \ldots, X_n$ be a random sample from a probability distribution with mean μ, variance σ^2, and moment generating function $M_X(t)$. Find the moment generating function for the sample mean $\overline{X} = \frac{1}{n} \sum_{i=1}^{n} X_i$ in terms of $M_X(t)$. Use this moment generating function to directly show that $E[\overline{X}] = \mu$ and $Var(\overline{X}) = \frac{\sigma^2}{n}$.

7.12. Let $X_1, \ldots, X_n$ be independent random variables with $X_i \sim n\left(\mu_i, \sigma_i^2\right), i = 1, \ldots, n$, and let $a_1, \ldots, a_n$ be arbitrary constants. Use the moment generating function technique to obtain the distribution of $V = \sum_{i=1}^{n} a_i X_i$.

7.13. Let $X_1, \ldots, X_n$ be independent random variables such that $X_i \sim Poisson\ (\lambda_i)$, with $\lambda_i > 0$, for $i = 1, \ldots, n$. Use the moment generating function technique to obtain the distribution of $U = \sum_{i=1}^{n} X_i$.

7.14. Let $X_1, \ldots, X_n$ be a random sample from the continuous probability distribution with p.d.f.:

$$f_X(x) = \theta x^{\theta-1} I_{(0,1)}(x), \text{ for } \theta > 0.$$

(a) Let $V_1 = -\ln(X_1)$. Find the probability distribution for V_1.

(b) Let $V_i = -\ln(X_i)$, for $i = 1, \ldots, n$. Find the probability distribution for $T = \sum_{i=1}^{n} V_i = \sum_{i=1}^{n} -\ln(X_i)$.

(c) Find $E\left[\frac{n}{T}\right]$. (Hint: You do *not* need to find the probability distribution of $\frac{1}{T}$.)

7.15. Consider the setting for Example 7.12 (logistic distribution). Use the general formula for the p.d.f. of the sample minimum, $X_{(1)}$, directly, without differentiating the associated c.d.f., to arrive at the same result as in that example.

7.16. Let $Y_1, \ldots, Y_n$ be mutually independent random variables such that $Y_i \sim Gamma\ (\alpha_i, \beta), i = 1, \ldots, n$. Use an induction argument to show that

$$W = \sum_{i=1}^{n} Y_i \sim Gamma\ (\alpha_1 + \cdots + \alpha_n, \beta).$$

7.17. Let $X_1,\ldots,X_n$ and $Y_1,\ldots,Y_n$ be mutually independent random variables such that $X_i \sim Gamma\,(\alpha_i, \beta)$, $i = 1, \ldots, m$, and $Y_j \sim Gamma\left(\alpha_j^*, \beta\right)$, for $j = 1, \ldots, n$. What is the probability distribution of the random variable

$$W = \frac{X_1 + \cdots + X_m}{X_1 + \cdots + X_m + Y_1 + \cdots + Y_n}?$$

Justify your answer.

7.18. Let $W \sim Beta\,(\alpha_1, \alpha_2)$. For any $k > 0$, show that

$$E\left[W^k\right] = \frac{\Gamma(\alpha_1 + \alpha_2)\Gamma(\alpha_1 + k)}{\Gamma(\alpha_1)\Gamma(\alpha_1 + \alpha_2 + k)}.$$

Use this result to ascertain that

$$E[W] = \frac{\alpha_1}{\alpha_1 + \alpha_2} \quad \text{and} \quad Var(W) = \frac{\alpha_1\alpha_2}{(\alpha_1 + \alpha_2 + 1)(\alpha_1 + \alpha_2)^2}.$$

7.19. Let U and V be random variables such that V and $\frac{U}{V}$ are independent. Show that $E\left[\frac{U}{V}\right] = \frac{E[U]}{E[V]}$, provided $E[V] \neq 0$.

7.20. Let $X_1,\ldots,X_n$ be independent random variables such that the distribution of X_i is symmetric about some value μ_i, for $i = 1, \ldots, n$. Show that

$$(X_1 - \mu_1, \ldots, X_n - \mu_n) \stackrel{d}{=} (\mu_1 - X_1, X_2 - \mu_2, \ldots, X_n - \mu_n) \stackrel{d}{=} \cdots$$

$$\stackrel{d}{=} (\mu_1 - X_1, \ldots, \mu_n - X_n),$$

where all 2^n such terms appear in this string of equalities in distribution.

7.21. Let $X_1,\ldots,X_n$ be a random sample from a distribution that is symmetric about the point μ, where n is an odd integer. Let $M = X_{\left(\frac{n+1}{2}\right)}$ be the sample median. Show that the distribution of M is also symmetric about the point μ, and, hence, that M is an unbiased estimator for μ in this setting, provided the expectation exists.

7.22. Consider the same setting as in Exercise 7.21, except now take n to be an even integer. Is the sample median M still an unbiased estimator for μ (provided the expectation exists)? Justify your answer.

7.23. Let $M = \text{median}(X_1,\ldots,X_n)$ be the sample median for a random sample $X_1,\ldots,X_n$. Show that M is an odd translation statistic.

7.24. Let $X_{(1)} \leq \cdots \leq X_{(n)}$ be the order statistics for the sample data $X_1,\ldots,X_n$. Show that the sample range $R = X_{(n)} - X_{(1)}$ is an even translation-invariant statistic.

7.25. Let $u(X_1,\ldots,X_n)$ and $v(X_1,\ldots,X_n)$ be two odd translation statistics. What property does the difference $t(X_1,\ldots,X_n) = u(X_1,\ldots,X_n) - v(X_1,\ldots,X_n)$ possess? Give an example of such a statistic $t(X_1,\ldots,X_n)$.

7.26. Construct an example of an even translation statistic.

7.27. Let $X_1,\ldots,X_n$ be a random sample from a distribution that is symmetric about μ. Argue that the sample midrange $(X_{(1)} + X_{(n)})/2$ and sample mode are unbiased estimators for μ, provided the expectations exist.

7.28. Let $X_1,\ldots,X_n$ be a random sample from a distribution that is symmetric about μ. Show that an odd translation-invariant statistic $V(X_1,\ldots,X_n)$ is symmetrically distributed about 0. What does this say about the difference between the sample mean and sample median, $\overline{X} - M$, for this setting?

7.29. Let $X_1,\ldots,X_n$ be a random sample from a distribution that is symmetric about μ. Each of the $(n(n + 1)/2)$ pairwise averages $U_{ij} = \frac{X_i+X_j}{2}$, for $i \leq j = 1, \ldots, n$, is called a *Walsh average*. Let $W = median\{ U_{ij}, i \leq j = 1,\ldots,n\}$. Show that W is an unbiased estimator for μ, provided the expectation exists.

7.30. Let $X_{(1)} \leq \ldots \leq X_{(n)}$ be the order statistics for a random sample $X_1,\ldots,X_n$ from a distribution that is symmetric about μ. Define

$$U = \sum_{i=1}^{n} a_i X_{(i)},$$

where $a_1, \ldots, a_n$ are constants satisfying

$$\sum_{i=1}^{n} a_i = 1 \text{ and } a_i = a_{n+1-i}, \ i = 1, \ldots, n.$$

Show that U has a distribution that is symmetric about μ, and, hence, is unbiased for μ if the expectation exists. Give an example of a statistic U that satisfies these conditions.

7.31. Prove Corollary 7.3. (Hint: Consider the function $g_1(\cdot)$ that maps $(x_1,\ldots,x_n)$ into $(2\mu - x_1,\ldots,2\mu - x_n)$.)

7.32. Let $X_{(1)} \leq \ldots \leq X_{(n)}$ be the order statistics for a random sample from a continuous distribution that is symmetric about μ. Define

$$U = \sum_{i=1}^{n} a_i X_{(i)} \text{ and } V = \sum_{j=1}^{n} b_j X_{(j)},$$

where $a_1, \ldots, a_n$ and $b_1, \ldots, b_n$ are constants that satisfy

$$\sum_{i=1}^{n} a_i = 1 \quad \text{and} \quad a_i = a_{n+1-i}, \quad i = 1, \ldots, n,$$

and

$$\sum_{j=1}^{n} b_j = 0 \quad \text{and} \quad b_j = -b_{n+1-j}, \quad j = 1, \ldots, n.$$

Show that $Cov(U, V) = 0$ provided it exists. Give examples of two statistics U and V that satisfy these conditions.

7.33. Let X be a continuous random variable with p.d.f.:

$$f_X(x) = \frac{1}{x\sqrt{2\pi\sigma^2}} e^{-(\ln(x)-\mu)^2/2\sigma^2} I_{(0,\infty)}(x) I_{(-\infty,\infty)}(\mu) I_{(0,\infty)}(\sigma^2).$$

(a) Find and identify the p.d.f. for $Y = \ln X$.

(b) Use the result in part (a) to deduce that $E[X] = e^{\mu + \frac{\sigma^2}{2}}$.

(c) Let $X_1, \ldots, X_n$ be a random sample of size n from $f_X(x)$. Find the sampling distribution of the statistic $Y = \ln\left(\prod_{i=1}^{n} X_i\right)$.

7.34. Let $X_{(1)} \leq \cdots \leq X_{(n)}$ be the order statistics for a random sample of size n from the continuous distribution with p.d.f.:

$$f_X(x) = e^{-(x-\theta)} I_{(\theta,\infty)}(x) I_{(-\infty,\infty)}(\theta).$$

Find and identify the sampling distribution of $W = n(X_{(1)} - \theta)$.

7.35. Consider a bowl containing four chips numbered 5, 5, 10, and 50, respectively. Draw three chips at random from the bowl without replacement between draws.

(a) Obtain the sampling distribution for the statistic $W = $ [largest number obtained $-$ smallest number obtained].

(b) Find $E[W]$ and $Var(W)$.

(c) Suppose we repeat this same experiment 100 independent times and let $V = $ [number of these 100 experiments for which $W = 5$]. Find an exact analytical expression for $P(20 \leq V \leq 30)$. (You do not need to obtain a numerical value for this probability, but you might try using **R** to evaluate it.)

(d) Obtain an approximate numerical value for the probability in part (c).

(e) Let Q denote the sum of the 100 values of W obtained in the 100 independent experiments from part (c). Find an approximate numerical value for $P(3475 \leq Q \leq 3550)$.

7.36. Let X be a continuous random variable with p.d.f.:

$$f_X(x) = e^{-(x-\theta)} I_{(\theta,\infty)}(x) I_{(-\infty,\infty)}(\theta).$$

(a) Find the moment generating function for $Y = X - \theta$ and use it to find the moment generating function for X.

(b) Let $X_1,\ldots,X_n$ be a random sample of size n from $f_X(x)$ and set $W = \sum_{i=1}^{n} X_i$. Find the moment generating function for $V = W - n\theta$ and identify the distribution for V.

(c) Use the result in part (b) and a change of variable to find the p.d.f. for W.

7.37. Let $X_{(1)} \le \cdots \le X_{(n)}$ be the order statistics for a random sample of size $n \ge 2$ from the continuous distribution with p.d.f.:

$$f_X(x) = \frac{1}{x^2} I_{(1,\infty)}(x).$$

(a) Find $P(X_{(n)} < 3)$.
(b) Obtain $E[X_{(n)}]$.

7.38. Consider the bowl containing chips numbered -5, 0, 2, 4, and 9. Draw three chips at random from the bowl without replacement between the draws. Construct the sampling distribution for the sample median $X_{(3)}$ and find $E[X_{(3)}]$ and $Var(X_{(3)})$.

7.39. Let $X_1,\ldots,X_n$ be a random sample of size n from the continuous distribution with p.d.f.:

$$f_X(x) = \frac{1}{\beta} e^{\frac{x}{\beta}} I_{(-\infty,0)}(x) I_{(0,\infty)}(\beta).$$

Find and identify the sampling distribution of the statistic $Y = -\sum_{i=1}^{n} X_i$.

7.40. Roll a pair of fair dice and let $Z =$ [larger number on the two dice—smaller number on the two dice].

(a) What is the probability distribution for Z?
(b) Find $E[Z]$ and $Var(Z)$.
(c) Flip a coin Z times and let $Y =$ [number of heads obtained on the Z flips]. Find $E[Y]$ and $Var(Y)$.

7.41. Let $X_1,\ldots,X_n$ be mutually independent random variables such that $X_i \sim Gamma\,(\alpha_i, \beta_i)$, with $\alpha_i > 0$ and $\beta_i > 0$ for $i = 1, \ldots, n$.

(a) Find and identify the probability distribution for $V = \sum_{i=1}^{n} \frac{X_i}{\beta_i}$.
(b) Find the moment generating function for $Q = \ln(X_i^{\alpha_i})$. (Do not expect to recognize this moment generating function.)

7.42. Let $X_{(1)} \leq \cdots \leq X_{(n)}$ be the order statistics for a random sample of size n from the $Unif(0, 1)$ distribution.

(a) Show that the joint p.d.f. of $(X_{(1)}, X_{(n)})$ is given by

$$f_{X_{(1)}, X_{(n)}}\left(x_{(1)}, x_{(n)}\right) = n(n-1)\left[x_{(n)} - x_{(1)}\right]^{n-2} I_{\{0 < x_{(1)} < x_{(n)} < 1\}}\left(x_{(1)}, x_{(n)}\right).$$

(b) Let $R = X_{(n)} - X_{(1)}$ be the sample range and set $V = X_{(n)}$. Find the joint p.d.f. for (R, V).
(c) Are R and V independent variables? Justify your answer.
(d) Find the marginal p.d.f.'s for R and V and identify the p.d.f. for R.

7.43. Let $X_{(1)} \leq \cdots \leq X_{(n)}$ be the order statistics for a random sample of size $n \geq 2$ from the continuous distribution with p.d.f.:

$$f_X(x) = \frac{1}{x^2} I_{(1,\infty)}(x).$$

Find $E[X_{(1)}]$.

7.44. Let X be a continuous random variable with p.d.f.:

$$f_X(x) = e^{-|x|} I_{(-\infty,\infty)}(x).$$

(a) Find the c.d.f. for $Y = |X|$ and identify the probability distribution.
(b) Now, let $X_1, \ldots, X_n$ be a random sample of size n from $f_X(x)$. What is the probability distribution for $V = \sum_{i=1}^{n} |X_i|$? Justify your answer.

7.45. Two persons, say A and B, separately roll a fair six-sided die until they each roll their first three. Let X denote the number of rolls required by A, and let Y denote the number of rolls required by B.

(a) Find $P(\text{maximum}(X, Y) < 10)$.
(b) Find $P(\text{minimum}(X, Y) < 10)$.
(c) Find $P(X + Y \text{ is an even number})$.

7.46. Let $X_{(1)} \leq \cdots \leq X_{(n)}$ be the order statistics for a random sample of size n from the continuous distribution with p.d.f.:

$$f_X(x) = 3x^2 I_{(0,1)}(x).$$

Find $E[X_{(n)}]$ and $Var(X_{(n)})$.

7.47. Let X and Y be independent geometric random variables with common parameter p, $0 < p < 1$. Use the change of variable approach to obtain the probability distribution of $W = X + Y$ and identify this distribution. Does it make intuitive sense?

7.48. Let X and Y be independent and identically distributed as *Gamma* (α, β) with $\alpha > 0$ and $\beta > 0$.

(a) Find the moment generating function for the random variable $W = Y - X$.
(b) Use the moment generating function from (a) to find $E[W^3]$.

7.49. Let X and Y be independent and identically distributed continuous random variables with common p.d.f.:

$$f_X(x) = \theta x^{\theta-1} \, I_{(0,1)}(x) I_{(0,\infty)}(\theta).$$

Find the p.d.f. for the variable $Z = XY$.

7.50. Let $X_1,\ldots,X_m$ and $Y_1, \ldots, Y_n$ be independent random samples of sizes m and n, respectively, from the same continuous distribution with c.d.f. $F(x)$ and p.d.f. $f(x)$. Let $X_{(1)}$ and $Y_{(n)}$ be the smallest X and largest Y, respectively. Find $P(Y_{(n)} < X_{(1)})$. Does this answer make intuitive sense?

7.51. Let $X_1,\ldots,X_n$ be a random sample from the *Geom* (p) distribution with $0 < p < 1$.

(a) Find the c.d.f. for this geometric distribution.
(b) Let $Q = \text{minimum } \{X_1,\ldots,X_n\}$. Find the c.d.f. for Q and identify the probability distribution for Q. Does this result make intuitive sense?

7.52. Let $X_1,\ldots,X_n$ be a random sample from the continuous distribution with p.d.f.

$$f_X(x) = e^{-(x-\theta)} \, I_{(\theta,\infty)}(x) \, I_{(-\infty,\infty)}(\theta).$$

(a) Find the c.d.f. for the X distribution.
(b) Set $U = \text{minimum } \{X_1,\ldots,X_n\}$. Find the c.d.f. for U.

7.53. Roll a pair of fair dice and let V be the absolute difference between the values on the two dice.

(a) Find the probability distribution for V.
(b) If we repeat this process of rolling the pair of dice ten times, what is the probability that we never obtain an absolute difference on the two die faces greater than 3?

7.54. Let W and Z be independent, continuous random variables with p.d.f.'s $g_W(w)$ and $h_Z(z)$, respectively.

(a) Show that

$$P(W < Z) = \int\limits_{-\infty}^{\infty} \int\limits_{-\infty}^{z} g_W(w) h_Z(z) \, dw \, dz.$$

(b) Let $X_1,\ldots,X_m$ and $Y_1, \ldots, Y_n$ be independent random samples of sizes m and n, respectively, from the same continuous distribution with c.d.f. $F(x)$ and p.d.f. $f(x)$. Find $P(X_{(m)} < Y_{(1)})$. Does this result make intuitive sense?

7.55. Suppose that a population has six men with weights of 110, 135, 150, 160, 170, and 190 pounds, respectively. Two men are sampled with replacement.

(a) Obtain the sampling distribution for the weight of the heaviest man in the sample.
(b) What is the probability that the maximum weight in the sample exceeds 170 pounds?
(c) What are the expected value and variance for the sample maximum weight?
(d) Suppose, instead, that the sampling is done with replacement. What is your new answer to part (b)? (Obtain this answer without finding the entire sampling distribution of the maximum weight under replacement.)

7.56. Let X be a continuous random variable with p.d.f.:

$$f_X(x) = \frac{1}{2}e^{-|x-\theta|}I_{(-\infty,\infty)}(x)I_{(-\infty,\infty)}(\theta)$$

(a) Find the c.d.f. for this probability distribution.
(b) Show that the probability distribution is symmetric about θ.
(c) Let $X_1,\ldots,X_n$ be a random samples of size n from this distribution, where n is an odd integer. Set $V = \sum_{i=1}^{n}(X_i - \widetilde{X})$, where $\widetilde{X} = X_{\left(\frac{n+1}{2}\right)}$ is the sample median. Show that the probability distribution for V is symmetric about 0.
(d) Find $E[\widetilde{X}]$.

7.57. Let (X,Y) be continuous variables with joint p.d.f.:

$$f_{X,Y}(x,y) = 2Ce^{-Cx}I_{\{0<y<x<\infty\}}(x,y),$$

where C is a constant.

(a) Find the value of C so that $f_{X,Y}(x,y)$ is a proper p.d.f.
(b) Find the p.d.f. for the random variable $Z = X - Y$.

7.58. Let X be a continuous random variable with p.d.f.:

$$f_X(x) = 2x(x^2 - 1)e^{-(x^2-1)}I_{(1,\infty)}(x).$$

(a) Find the p.d.f. of $Z = X^2 - 1$. Identify the probability distribution.
(b) Suppose X_1, X_2, X_3 is a random sample of size 3 from $f_X(x)$. Find and identify the probability distribution for the variable $W = X_1^2 + X_2^2 + X_3^2 - 3$.

7.59. Let (X, Y) be continuous random variables with joint p.d.f.:

$$f_{X,Y}(x, y) = 8(y - x)e^{-2y}I_{\{0<x<y<\infty\}}(x, y).$$

(a) Define $W = Y + X$. Find the p.d.f. for W.
(b) Derive the moment generating function for $Z = Y - X$ without first finding the p.d.f. for Z.

7.60. Let $X \sim Exp$ (1) and $Y \sim Exp$ (1) be independent random variables. Find the p.d.f. for the variable $W = \ln\left(\frac{X}{Y}\right)$.

7.61. Let $X \sim Exp(\lambda_X)$ and $Y \sim Exp(\lambda_Y)$ be independent random variables. Find the p.d.f. for the variable $Z = \frac{\lambda_X X}{\lambda_Y Y}$.

7.62. Let $X \sim Exp$ (1) and $Y \sim Exp$ (1) be independent random variables. Find the p.d.f. for the variable $W = X - Y$.

7.63. Let $X_1, \ldots, X_n$ be independent random variables such that $X_i \sim Exp(\lambda_i)$, for $i = 1, \ldots, n$. Find and identify the p.d.f. for the random variable $W = minimum$ $\{X_1, \ldots, X_n\}$.

7.64. Let $X_1, \ldots, X_n$ be independent random variables such that $X_i \sim Exp(\lambda_i)$, for $i = 1, \ldots, n$. Find the c.d.f. and p.d.f. for the random variable $W = maximum$ $\{X_1, \ldots, X_n\}$.

7.65. Let $X_{(1)} \leq X_{(2)}$ be the order statistics for a random sample of size 2 from the continuous distribution with p.d.f.:

$$f_X(x) = I_{(\theta-\frac{1}{2}, \theta+\frac{1}{2})}(x)I_{(-\infty,\infty)}.(\theta)$$

(a) What is the joint p.d.f. for the pair $(X_{(1)}, X_{(2)})$?
(b) Compute $P(X_{(1)} < \theta < X_{(2)})$.
(c) Compute $P\left(X_{(1)} - \frac{1}{2} < \theta < X_{(2)} + \frac{1}{2}\right)$.
(d) Compute $P\left(X_{(1)} < \theta < X_{(2)} \middle| (X_{(2)} - X_{(1)}) > \frac{1}{2}\right)$.

7.66. Samantha and Hector are playing a game. Each player generates independent random numbers according to an exponential distribution with expectation β. Samantha generates three numbers, whose ordered values are represented by $X_{(1)} \leq X_{(2)} \leq X_{(3)}$. Hector generates only one number, say Y. The player with the highest number wins the game.

(a) Find the c.d.f. for $X_{(3)}$.
(b) Find the p.d.f. for $X_{(2)}$.
(c) What is the probability that Samantha wins the game?
(d) Find the p.d.f. for the difference $X_{(3)} - Y$.
(e) Find the moment generating function for $X_{(3)}$, being sure to specify the domain for the existence of the m.g.f.
(f) Find $E[X_{(3)}]$.

7.67. Let (X, Y) be random variables with the joint probability distribution

$$f_{X,Y}(x, y) = k \binom{n}{x} y^{x+\alpha-1} (1 - y)^{n-x+\beta-1} I_{\{0,1,\ldots,n\}}(x) I_{(0,1)}(y),$$

where n is a positive integer, $\alpha > 0$, and $\beta > 0$, and k is the normalizing constant to make $f_{X,Y}(x, y)$ a proper probability distribution. [Note that this is a mixed probability distribution with X being a discrete variable and Y being a continuous variable.]

(a) Determine the constant k.
(b) Derive and identify the marginal distribution of Y.
(c) Find and identify the conditional distribution of X given $Y = y \in (0, 1)$.
(d) Are X and Y independent? Justify your answer.
(e) Derive the marginal probability distribution for X.
(f) Find $E[X]$.

7.68. Let (X_i, Y_i), $i = 1, \ldots, n$, be mutually independent random variables such that $X_1, \ldots, X_n$ is a random sample from a *Poisson* (λ) distribution, with $\lambda > 0$, and $Y_1, \ldots, Y_n$ is a random sample from a Bernoulli distribution with parameter π, $0 \leq \pi \leq 1$. Define $Z_i = (2Y_i - 1)X_i$, for $i = 1, \ldots, n$.

(a) Find the mean and variance for the random variable Z_i.
(b) Find the joint probability function for $(Z_1, \ldots, Z_n)$.
(c) Define the random variables

$$N^+ = \sum_{i=1}^{n} I_{\{Z_i > 0\}} \text{ and } N^- = \sum_{i=1}^{n} I_{\{Z_i < 0\}}.$$

What is the probability distribution of $W = N^+ + N^-$?
(d) Find the conditional expected value $E[N^+ | W = j]$, for $j = 1, \ldots, n$.

7.69. Let $X_1, \ldots, X_{100}$ be a random sample of size $n = 100$ from the $Unif(\theta - 5, \theta + 5)$ distribution with $-\infty < \theta < \infty$, and consider the random intervals

$$I_i(X_i) = [X_i - 1, X_i + 2] \text{ for } i = 1, \ldots, 100.$$

(a) What is the probability that the interval $I_i(X_i)$ will contain the true parameter value θ?
(b) Let M denote the number of the $I_i(X_i)$ intervals that will *not* cover the true parameter value θ. Calculate the expected value of M and its variance.
(c) Of those $I_i(X_i)$ intervals that do *not* cover the true parameter value θ, some will fall entirely to the right of θ, and some will fall entirely to its left. Let R denote the number of $I_i(X_i)$ intervals that fall entirely to the right of the true parameter value θ.

   (i) Calculate the probability that an interval $I_i(X_i)$ fall entirely to the right of the true parameter value θ, given that it does not cover it.

(ii) Calculate the expected value and variance of R.

(d) Suppose you observe $x_7 = 1/(\sqrt{2\pi}2.5)$. What is the probability that the interval $I_7(x_7)$ falls entirely to the right of the true parameter value θ?

7.70. Let X_1, X_2 be a random sample of size 2 from the $Unif(0, 1)$ distribution.

(a) Set $Y_1 = \sqrt{-2\ln(X_1)}\cos(2\pi X_2)$ and $Y_2 = \sqrt{-2\ln(X_1)}\sin(2\pi X_2)$. Find the joint p.d.f. of (Y_1, Y_2).
(b) Are Y_1 and Y_2 independent? Justify your answer.
(c) Identify the marginal distributions of Y_1 and Y_2.
(d) Based on your solutions to (a) – (c), outline an algorithm to obtain a random sample of size 2 from a normal distribution with mean μ and variance σ^2.

7.71. Let $X_{(1)} \leq \cdots \leq X_{(n)}$ be the order statistics for a random sample of size n from the continuous distribution with p.d.f.:

$$f_X(x) = \frac{x}{\alpha} e^{-x^2/2\alpha} I_{(0,\infty)}(x) I_{(0,\infty)}(\alpha).$$

(This is called the *Rayleigh distribution with parameter* $\alpha > 0$.) Find and identify the probability distribution of $X_{(1)}$.

7.72. Let X and Y be two arbitrary, but independent, random variables with p.d.f.'s $f_X(x)$ and $f_Y(y)$, means μ_X and μ_Y, and variances σ_X^2 and σ_Y^2, respectively. Let $0 \leq p \leq 1$ be a constant. Consider the following two distinct ways to "mix" these two random variables:

$$A = pX + (1-p)Y \text{ (direct mixing)}$$
$$B \sim f_B(b) = p\,f_X(b) + (1-p)f_Y(b) \text{ (mixing of densities)}.$$

(a) Find $E[A]$, $Var(A)$, $E[B]$, and $Var(B)$ as functions of the means μ_X, μ_Y and variances σ_X^2, σ_Y^2.
(b) Show that $Var(A) \leq Var(B)$, with strict inequality unless $p = 0$ or 1. Discuss the implication of this result.
(c) Now suppose that $X \sim n(-4, 9)$ and $Y \sim n(4, 9)$. Are the associated random variables A and B also normally distributed? Justify your answer.

7.73. Let A be the area of a circle with radius $R \sim Exp(\beta)$.

(a) Find the p.d.f. for A.
(b) Let $A_1, \ldots, A_n$ be a random sample from the p.d.f. for A obtained in part (a). One estimator for the parameter β (the *maximum likelihood estimator*) is given by

$$\hat{\beta} = \frac{1}{n}\sum_{i=1}^{n}\left(\frac{A_i}{\pi}\right)^{\frac{1}{2}}.$$

Find $E\left[\hat{\beta}\right]$ and $Var\left(\hat{\beta}\right)$.

7.74. Let $T_i \sim Exp(\beta_i)$, $i = 1, ..., n$, be independent random variables.

(a) Find $P(T_i < T_j)$ for $i, j \in \{1, ..., n\}$, $i \neq j$.
(b) Let $T_{(1)} \leq \cdots \leq T_{(n)}$ be the ordered values for $T_1, ..., T_n$. Find the c.d.f. and p.d.f. for both $T_{(1)}$ and $T_{(n)}$.
(c) Obtain $P(T_{(1)} < T_{(n)})$.
(d) Define K to be the index of the random variable that achieves the minimum; that is, K is such that $T_K = T_{(1)}$. Find the probability distribution for K. (Note that K is a discrete random variable.)

7.75. Suppose that $\{\epsilon_t : t = 0, 1, ...\}$ is a sequence of independent and identically distributed random variables with mean 0 and variance σ^2. The sequence $\{X_t : t = 1, ..., n\}$ is generated according to the following model:

$$X_t = \epsilon_t - \theta\,\epsilon_{t-1},$$

for some $\theta > 0$. Let $\overline{X} = \frac{1}{n}\sum_{i=1}^{n} X_i$ and $S^2 = \frac{1}{n-1}\sum_{i=1}^{n}(X_i - \overline{X})^2$.

(a) Find $E[X_t]$ and $Var(X_t)$, for $t = 1, ..., n$.
(b) Calculate $E[\overline{X}]$ and $Var(\overline{X})$.
(c) Find $E[S^2]$.
(d) Is S^2 an unbiased estimator for the variance of X_t or does it overestimate or underestimate it on average when $\theta < 0$?

7.76. Let $X_1, ..., X_n$ be a random sample of size n from the *Binom* (m, p) distribution where m is a known positive integer and $0 < p < 1$.

(a) Construct an unbiased estimator for $h(p) = (1 - p)^m$.
(b) What is the probability distribution for $T = \sum_{i=1}^{n} X_i$?
(c) Find the conditional distribution of T given $X_1 = 0$.

7.77. Let $X \sim Unif(-\theta, \theta)$ for some $\theta > 0$.

(a) What is the probability distribution of $Y = |X|$?
    Let $X_1, ..., X_n$ be a random sample of size n from the *Unif* $(-\theta, \theta)$ distribution and set $T = C \max\{|X_1|, ..., |X_n|\}$, where C is a constant that may depend on n but not on θ or $X_1, ..., X_n$
(b) Determine the value of C such that T is an unbiased estimator for θ.
(c) Find the variance of the unbiased estimator for θ obtained in part (b).
(d) Consider a second possible estimator for θ given by

$$W = \frac{2}{n}\sum_{i=1}^{n} |X_i|.$$

Show that W is also an unbiased estimator for θ.
(e) Which of the two unbiased estimators, T from part (b) or W from part (d), do you prefer? Justify your answer.

7.78. Let X_1, X_2, X_3 be a random sample of size 3 from the *Exp* (β) distribution.

(a) Calculate the p.d.f. for the sample median $\widetilde{X}$.
(b) Construct a function of $\widetilde{X}$ that is an unbiased estimator for β.

7.79. Let $X_{(1)} \leq \cdots \leq X_{(n)}$ be the order statistics for a random sample of size n from the *Exp* (1) distribution. The **n sample spacings** are defined as follows:

$$U_j = X_{(j)} - X_{(j-1)}, \quad \text{for } j = 1, \ldots, n,$$

where we define $X_{(0)} \equiv 0$. Consider the weighted sample spacings given by

$$V_j = (n - j + 1)U_j, \quad j = 1, \ldots, n.$$

(a) Show that $V_1 = n\,X_{(1)} \sim Exp$ (1).
(b) Show that $V_j \sim Exp$ (1) for all $j \in \{2, \ldots, n\}$ as well.
(c) Consider the special case of $n = 2$. Use a change of variable technique to show the unusual property that $V_1 = n\,X_{(1)}$ and $V_2 = X_{(2)} - X_{(1)}$ are independent variables.

[*Note:* It is even more unusual than this special case, as $V_1, \ldots, V_n$ are all mutually independent for an arbitrary positive integer n and, as you found in parts (a) and (b), each of them is distributed as *Exp* (1)!! If you are feeling brave, you can try an n-dimensional change of variable approach to prove this more general result for an arbitrary n.]
(d) Express the largest order statistic $X_{(n)}$ as a function of the n weighted spacings $V_1, \ldots, V_n$. Use the general independence of $V_1, \ldots, V_n$ as noted in part (c) to obtain expressions for $E[X_{(n)}]$ and $Var(X_{(n)})$. [**Think About It:** Perhaps you would prefer to simply obtain $E[X_{(n)}]$ and $Var(X_{(n)})$ directly from the p.d.f. for $X_{(n)}$. Give it a try!]

7.80. Assume that the number of typos on each page of an n-page textbook has a *Poisson* (λ) distribution. Also assume that the number of typos on different pages is independent. Let $X_1, \ldots, X_n$ denote the number of typos on pages 1, ..., n, respectively. Suppose we are interested in estimating the probability, γ, that a given page has no typos.

(a) Provide an expression for γ as a function of λ.
(b) Let $\overline{X} = \frac{1}{n} \sum_{i=1}^{n} X_i$ and set $T_n = e^{-\overline{X}}$. Show that T_n is not an unbiased estimator for γ.
(c) Let $W_n = [\#X's = 0]/n$. Show that W_n is an unbiased estimator for γ.
(d) Find $\widetilde{W} = E[W_n | X_1 + \cdots + X_n]$. Show that $\widetilde{W}$ is also an unbiased estimator for γ.
(e) Argue that $\widetilde{W}$ has smaller variance than W_n.

7.81. Let X_1, X_2 be a random sample of size 2 from a n (0, 1) distribution and set $Y = \sigma \sqrt{X_1^2 + X_2^2}$, where σ is a constant. Find the p.d.f. for Y.

7.82. Let $X_1, \ldots, X_n$ be a random sample of size n from a continuous distribution with p.d.f.:

$$f_X(x) = \theta(1 + x)^{-(1+\theta)} I_{(0,\infty)}(x) I_{(2,\infty)}(\theta).$$

Show that $T = \frac{1}{n} \sum_{i=1}^{n} \ln(1 + X_i)$ is an unbiased estimator for $\tau(\theta) = \frac{1}{\theta}$.

7.83. Let $X_1, \ldots, X_n$ be a random sample from a *Geom* (p) distribution. One commonly used estimator for the parameter p is $\hat{p} = \frac{1}{\overline{X}}$, where $\overline{X} = \frac{1}{n} \sum_{i=1}^{n} X_i$ is the sample mean. Show that $\hat{p}$ is not an unbiased estimator for p. [Hint: Consider the case of $n = 1$.]

7.84. Let $X_{(1)} \leq \cdots \leq X_{(n)}$ be the order statistics for a random sample of size $n \geq 2$ from the continuous distribution with p.d.f.:

$$f_X(x) = I_{(\alpha,\beta)}(x) I_{\{-\infty < \alpha < \beta < \infty\}}(\alpha, \beta).$$

(a) Find the p.d.f. for $X_{(1)}$.
(b) Obtain $E[X_{(1)}]$ and $Var(X_{(1)})$. [Hint: Work first with the variable $W_n = \beta - X_{(1)}$.]

7.85. Let $X_{(1)} \leq \cdots \leq X_{(n)}$ be the order statistics for a random sample of size $n \geq 2$ from the continuous distribution with p.d.f.:

$$f_X(x) = I_{(\alpha,\beta)}(x) I_{\{-\infty < \alpha < \beta < \infty\}}(\alpha, \beta).$$

(a) Find the p.d.f. for $X_{(n)}$.
(b) Obtain $E[X_{(n)}]$ and $Var(X_{(n)})$. [Hint: Work first with the variable $V_n = X_{(n)} - \alpha$.]

7.86. Let $X_{(1)} \leq \cdots \leq X_{(n)}$ be the order statistics for a random sample of size n from the continuous distribution with p.d.f. $f_X(x)$ and c.d.f. $F_X(x)$. It can be shown (you need not!) that the joint p.d.f. for $(X_{(1)}, \ldots, X_{(n)})$ is given by

$$f_{X_{(1)}, \ldots, X_{(n)}}(x_{(1)}, \ldots, x_{(n)}) = n! f_X(x_{(1)}) f_X(x_{(2)}) \cdots f_X(x_{(n)}) I_{\{-\infty < x_{(1)} < x_{(2)} < \cdots < x_{(n)} < \infty\}}(x_{(1)}, \ldots, x_{(n)}).$$

[**Think About It**: Does this joint distribution make intuitive sense?]

(a) Carefully integrating out the other $(n - 1)$ variables, show that the marginal distribution of $X_{(1)}$ is as obtained in Example 7.10 via the distribution function technique.
(b) Carefully integrating out the other $(n - 1)$ variables, show that the marginal distribution of $X_{(n)}$ is as obtained in Example 7.10 via the distribution function technique.
(c) Following the pattern observed in establishing parts (a) and (b), show that the marginal distribution of an arbitrary order statistic $X_{(j)}$, $j \in \{1, \ldots, n\}$, is as obtained in Example 7.13 via the distribution function technique.

Think About It: Could this same approach be used to obtain the joint distribution of two order statistics $(X_{(i)}, X_{(j)})$, $(i,j) \in \{1,\ldots,n\}$ with $i < j$? Go ahead—give it a try!

7.87. Let $X_{(1)} \leq \cdots \leq X_{(n)}$ be the order statistics for a random sample of size n from the *Exp* (1) distribution.

(a) Find the form of the p.d.f. for $X_{(j)}, j \in \{1,\ldots,n\}$, and use it to obtain $E[1 - e^{-X_{(j)}}]$

   and $Var(1 - e^{-X_{(j)}})$. [Hint: Consider the change of variable $Y = 1 - e^{-X_{(j)}}$ in the relevant integrals.]

(b) Let $X \sim Exp$ (1). What is the probability distribution for $W = 1 - e^{-X}$? [Hint: You do not need to actually make the change of variable to determine the answer.]

(c) Can the result in part (b) be used to argue for the values of $E[1 - e^{-X_{(j)}}]$ and $Var(1 - e^{-X_{(j)}})$ obtained in part (a) without directly considering the p.d.f. for $X_{(j)}$?

   Justify your answer.

7.88. Let X and Y be random variables with means μ_X and μ_Y, respectively, and variances σ_X^2 and σ_Y^2, respectively.

(a) Show that $E[X + Y] = E[X] + E[Y]$.
(b) If, in addition, X and Y are independent variables, show that $Var(X + Y) = \sigma_X^2 + \sigma_Y^2$.

7.89. Let $X_1,\ldots,X_n$ be random variables with means $\mu_{X_1}, \ldots, \mu_{X_n}$ and variances $\sigma_{X_1}^2, \ldots, \sigma_{X_n}^2$, respectively. Use the results in Exercise 7.88 and Mathematical Induction to show that:

(a) $E[X_1 + \cdots + X_n] = \mu_{X_1} + \cdots + \mu_{X_n}$.
(b) $Var(X_1 + \cdots + X_n) = \sigma_{X_1}^2 + \cdots + \sigma_{X_n}^2$, if $X_1,\ldots,X_n$ are also mutually independent variables.

7.90. Let $X_1,\ldots,X_n$ be a random sample of size n from a probability distribution with mean μ and variance σ^2, and let $\overline{X} = \frac{1}{n}\sum_{i=1}^{n} X_i$ be the sample mean. Show that $E[\overline{X}] = \mu$ and $Var(\overline{X}) = \frac{\sigma^2}{n}$.

7.91. Let (X, Y) be continuous random variables with joint p.d.f.:

$$f_{X,Y}(x, y) = \frac{1}{4!6^6} y^3 e^{-\frac{x}{6} - \frac{y}{x}} I_{(0,\infty)}(x) I_{(0,\infty)}(y)$$

(a) Find the joint p.d.f. of $W = Y/X$ and $V = X$.
(b) Are W and V independent variables? Justify your answer.
(c) Find and identify the marginal p.d.f.'s for W and V.
(d) Find $E[Y]$ without finding the marginal distribution for Y.

Chapter 8
Asymptotic (Large-Sample) Properties of Statistics

Quite often in statistical applications it is important to know how relevant statistics and their sampling distributions behave as the sample size(s) n increases (i.e., goes to infinity). Such properties are referred to as *asymptotic* or *large-sample* ($n \to \infty$) *properties*. For our purposes in this text, we concentrate on two of the most basic asymptotic properties: convergence in probability and convergence in distribution.

8.1 Convergence in Probability

Definition 8.1 Let $\{Z_n\}$ be a sequence of random variables such that the probability distribution for Z_n depends on the index n (usually the sample size). We say that Z_n **converges in probability to the constant α (as $n \to \infty$)** if

$$\lim_{n \to \infty} P(|Z_n - \alpha| > \varepsilon) = 0 \quad \text{for every } \varepsilon > 0.$$

We write $Z_n \xrightarrow{P} \alpha$. (Note that it is equivalent to show that $\lim_{n \to \infty} P(|Z_n - \alpha| \le \varepsilon) = 1$ for every $\varepsilon > 0$.)

Example 8.1 Sample Mean from a Normal Distribution

Let $X_1, \ldots, X_n$ be a random sample from a $n(\mu, \sigma^2)$ distribution and set $\overline{X} = \frac{1}{n} \sum_{i=1}^{n} X_i$. Then we know that $\overline{X} \sim n\left(\mu, \frac{\sigma^2}{n}\right)$. Thus, for any $\varepsilon > 0$

© Springer Nature Switzerland AG 2020
D. Wolfe, G. Schneider, *Primer for Data Analytics and Graduate Study in Statistics*,
https://doi.org/10.1007/978-3-030-47479-9_8

$$P(|\bar{X} - \mu| \le \varepsilon) = P(\mu - \varepsilon \le \bar{X} \le \mu + \varepsilon) = P\left(-\frac{\varepsilon}{\frac{\sigma}{\sqrt{n}}} \le \frac{\bar{X} - \mu}{\frac{\sigma}{\sqrt{n}}} \le \frac{\varepsilon}{\frac{\sigma}{\sqrt{n}}}\right)$$

$$= \Phi\left(\frac{\sqrt{n}\varepsilon}{\sigma}\right) - \Phi\left(-\frac{\sqrt{n}\varepsilon}{\sigma}\right),$$

where $\Phi(\cdot)$ is the c.d.f. for the standard normal $n\,(0, 1)$ distribution.

$$\Rightarrow \lim_{n \to \infty} P(|\bar{X} - \mu| \le \varepsilon) = \lim_{n \to \infty} \left[\Phi\left(\frac{\sqrt{n}\varepsilon}{\sigma}\right) - \Phi\left(-\frac{\sqrt{n}\varepsilon}{\sigma}\right)\right] = 1 - 0 = 1$$

$\Rightarrow \bar{X} \xrightarrow[n \to \infty]{P} \mu$, directly by the definition of convergence in probability.

Definition 8.2 Let $S = S(X_1, \ldots, X_n)$ be a statistic based on a random sample $X_1, \ldots, X_n$ from a probability distribution with parameter θ. We say that S is a **consistent estimator for the parameter θ** if S converges in probability to θ as $n \to \infty$.

Thus, we see from Example 8.1 that the sample mean, $\bar{X}$, is a consistent estimator for the population mean μ for a random sample from a normal distribution.

We seldom have to use the definition of convergence in probability directly to establish this property. There are many general results that make our life much easier than that. The first of these deals only with the mean and variance sequences for the sequence of random variables $\{Z_n\}$.

Theorem 8.1 Let $\{Z_n\}$ be a sequence of random variables with associated (finite) mean and variance sequences $\{\mu_n\}$ and $\{\sigma_n^2\}$, respectively, where $\mu_n = E[Z_n]$ and $\sigma_n^2 = Var(Z_n)$. If $\lim_{n \to \infty} \mu_n = \mu$ and $\lim_{n \to \infty} \sigma_n^2 = 0$, then $Z_n \xrightarrow[n \to \infty]{P} \mu$.

Proof For any $\varepsilon > 0$, we have

$$\{|Z_n - \mu| > \varepsilon\} = \{|Z_n \pm \mu_n - \mu| > \varepsilon\} \subset \{|Z_n - \mu_n| > \varepsilon/2\} \cup \{|\mu_n - \mu| > \varepsilon/2\},$$

since $|s + t| \le |s| + |t|$ for all s and t. Now, since $\lim_{n \to \infty} \mu_n = \mu$, we can choose an N^* such that

$$|\mu_n - \mu| \le \varepsilon/2 \quad \text{for all } n \ge N^*,$$

so that, for $n \ge N^*$, we have that $\{|\mu_n - \mu| > \varepsilon / 2\} = \emptyset$, the empty set. It follows that

$$P(|Z_n - \mu| > \varepsilon) \le P(|Z_n - \mu_n| > \varepsilon/2) \quad \text{for all } n \ge N^*.$$

Remembering that $\mu_n = E[Z_n]$, it follows from Chebyshev's Inequality (Theorem 4. 2) that

$$P(|Z_n - \mu_n| > \varepsilon/2) \leq \frac{Var(Z_n)}{(\varepsilon/2)^2} \leq \frac{4\sigma_n^2}{\varepsilon^2} \to 0 \quad \text{as } n \to \infty,$$

since $\lim\limits_{n\to\infty} \sigma_n^2 = 0$. Thus, we have

$$0 \leq \lim_{n\to\infty} P(|Z_n - \mu| > \varepsilon) \leq \lim_{n\to\infty} P(|Z_n - \mu_n| > \varepsilon/2) = 0$$

$$\Rightarrow \lim_{n\to\infty} P(|Z_n - \mu| > \varepsilon) = 0,$$

which yields the result that $Z_n \overset{P}{\to} \mu$, as $n \to \infty$. ∎

Note that if a statistic Z_n satisfies the conditions of Theorem 8.1, then it is a consistent estimator for the parameter μ.

Example 8.2 Sample Mean from a Normal Distribution

Consider again the setting of Example 8.1. Taking $Z_n = \overline{X}$, $\mu_n = \mu$, and $\sigma_n^2 = \frac{\sigma^2}{n}$, the consistency of the sample mean, $\overline{X}$, for a random sample from a normal distribution follows immediately from Theorem 8.1.

Example 8.3 Maximum Order Statistic for a Uniform Distribution

Let $X_{(1)} \leq \ldots \leq X_{(n)}$ be the order statistics for a random sample of size n from the $Unif(0, \theta)$ distribution, with $\theta > 0$. The p.d.f. for $X_{(n)}$ is given by

$$g_{X_{(n)}}(x) = n\frac{x^{n-1}}{\theta^n} I_{(0,\theta)}(x).$$

Thus,

$$E[X_{(n)}] = \int_0^\theta \frac{nx^n}{\theta^n}\, dx = \frac{nx^{n+1}}{(n+1)\theta^n}\bigg|_{x=0}^{x=\theta} = \frac{n\theta}{n+1}$$

and

$$E[X_{(n)}^2] = \int_0^\theta \frac{nx^{n+1}}{\theta^n}\, dx = \frac{nx^{n+2}}{(n+2)\theta^n}\bigg|_{x=0}^{x=\theta} = \frac{n\theta^2}{n+2},$$

which leads to

$$Var(X_{(n)}) = \frac{n\theta^2}{n+2} - \left[\frac{n\theta}{n+1}\right]^2 = \frac{n\theta^2}{(n+1)^2(n+2)}.$$

Since

$$\lim_{n\to\infty} E\big[X_{(n)}\big] = \lim_{n\to\infty} \frac{n\theta}{n+1} = \theta$$

and

$$\lim_{n\to\infty} Var\big(X_{(n)}\big) = \lim_{n\to\infty} \frac{n\theta^2}{(n+1)^2(n+2)} = 0,$$

it follows from Theorem 8.1 that $X_{(n)} \xrightarrow{P} \theta$, as $n \to \infty$, so that $X_{(n)}$ is a consistent estimator for θ.

Theorem 8.2 (Weak Law of Large Numbers—WLLN) Let $X_1, \ldots, X_n$ be a random sample of size n from a probability distribution with mean μ and finite variance σ^2. Let $\overline{X} = \frac{1}{n}\sum_{i=1}^{n} X_i$ be the sample mean. Then, $\overline{X} \xrightarrow{P} \mu$ as $n \to \infty$.

Proof We know that $E\big[\overline{X}\big] = \mu$ and $Var\big(\overline{X}\big) = \frac{\sigma^2}{n}$. The result then follows immediately from Theorem 8.1. ∎

Thus, the sample mean $\overline{X}$ is a consistent estimator for the population mean for *any* probability distribution with a finite variance. Note that this generalizes the result that we previously obtained in Example 8.1 by direct use (the hard way!) of the definition of convergence in probability specifically for the normal distribution.

The next result provides an important way to show convergence in probability for even complicated functions of a random variable.

Theorem 8.3 Suppose that $Z_n \xrightarrow[n\to\infty]{P} c$. If $g(\cdot)$ is a function that does not depend on n and is continuous at c, then $g(Z_n) \xrightarrow[n\to\infty]{P} g(c)$.

Proof Let $\varepsilon > 0$ be arbitrary. Since $g(t)$ is continuous at $t = c$, there exists a $\delta > 0$ such that

$$|t - c| < \delta \Rightarrow |g(t) - g(c)| < \varepsilon.$$

This, in turn, implies that

$$P(|Z_n - c| < \delta) \leq P(|g(Z_n) - g(c)| < \varepsilon) \leq 1 \quad \text{for all } n.$$

Combining this with our assumption that $Z_n \xrightarrow[n\to\infty]{P} c$, it follows from the definition of convergence in probability that

$$1 = \lim_{n \to \infty} P(|Z_n - c| < \delta) \leq \lim_{n \to \infty} P(|g(Z_n) - g(c)| < \varepsilon) \leq 1$$

$$\Rightarrow \lim_{n \to \infty} P(|g(Z_n) - g(c)| < \varepsilon) = 1 \ \Rightarrow \ g(Z_n) \xrightarrow[n \to \infty]{P} g(c). \ \blacksquare$$

Example 8.4 Distribution Function for a Normal Distribution

Let $X_1, \ldots, X_n$ be a random sample from a $n(\mu, 1)$ distribution and set $\overline{X} = \frac{1}{n} \sum_{i=1}^{n} X_i$. The c.d.f. for X is given by

$$F_X(t) = P(X \leq t) = \Phi\left(\frac{t - \mu}{\sqrt{1}}\right) = \Phi(t - \mu),$$

where $\Phi(\cdot)$ is the c.d.f. for the standard normal distribution. Now we know that (a) $\overline{X} \xrightarrow[n \to \infty]{P} \mu$ (WLLN) and (b) $\Phi(\cdot)$ is a continuous function everywhere. A natural estimator for $F_X(t)$ is $\widehat{F}_X(t) = \Phi(t - \overline{X})$, and it follows from Theorem 8.3 that $\widehat{F}_X(t) \xrightarrow[n \to \infty]{P} F_X(t)$ for every t; that is, $\widehat{F}_X(t) = \Phi(t - \overline{X})$ is a consistent estimator of $F_X(t)$ for every t.

Example 8.5 Continuous Distribution

Let $X_1, \ldots, X_n$ be a random sample from the continuous distribution with p.d.f.

$$f(x|\theta) = \theta x^{\theta - 1} I_{(0,1)}(x) I_{(0,\infty)}(\theta).$$

Note that

$$E[X] = \int_0^1 x \, \theta x^{\theta - 1} dx = \frac{\theta}{\theta + 1}.$$

Now the sample mean $\overline{X} = \frac{1}{n} \sum_{i=1}^{n} X_i$ is a natural estimator for $E[X]$, and we know from the WLLN that $\overline{X} \xrightarrow[n \to \infty]{P} E[X] = \frac{\theta}{1+\theta}$. Now let $g(c) = \frac{c}{1-c}$, which is a continuous function of c except at $c = 1$. From Theorem 8.3 (since $\frac{\theta}{1+\theta} \neq 1$), it follows that

$$g(\overline{X}) = \frac{\overline{X}}{1 - \overline{X}} \xrightarrow[n \to \infty]{P} \frac{\frac{\theta}{1+\theta}}{1 - \frac{\theta}{1+\theta}} = \theta.$$

Hence, $\widehat{\theta}_1 = \frac{\overline{X}}{1 - \overline{X}}$ is a reasonable sample-based estimator for θ. ($\widehat{\theta}_1$ is called the *method of moments estimator* for θ, and this example shows that it is a consistent estimator.)

We now state without proof a number of important properties of convergence in probability.

Theorem 8.4 Suppose that two random variables U_n and V_n are such that $U_n \xrightarrow[n \to \infty]{P} c$ and $V_n \xrightarrow[n \to \infty]{P} d$ for constants c and d. Then, it follows that

(a) $U_n + V_n \xrightarrow[n \to \infty]{P} c + d$,

(b) $U_n V_n \xrightarrow[n \to \infty]{P} cd$,

and

(c) $\frac{U_n}{V_n} \xrightarrow[n \to \infty]{P} \frac{c}{d}$, provided $d \neq 0$.

Proof Exercise 8.6 ∎

Example 8.6 Consistency of Sample Variance
Let $X_1, \ldots, X_n$ be a random sample from a probability distribution with mean μ, variance σ^2, and $E[X^4] < \infty$. Let $U_n = \frac{1}{n} \sum_{j=1}^{n} X_j^2$ and $V_n = -\overline{X}^2$. Then we have:

(i) $U_n \xrightarrow[n \to \infty]{P} E[X_1^2]$ by the *WLLN* (since $E[X^4] < \infty$)

(ii) $\overline{X} \xrightarrow[n \to \infty]{P} \mu$, also by the *WLLN* $\Rightarrow -\overline{X}^2 \xrightarrow[n \to \infty]{P} -\mu^2$ (taking $g(c) = -c^2$ in Theorem 8.3)

Now, (i) and (ii), in conjunction with part (a) in Theorem 8.4, imply that

$$S_n^2 = \frac{1}{n} \sum_{j=1}^{n} (X_j - \overline{X})^2 = \frac{1}{n} \sum_{j=1}^{n} X_j^2 - \overline{X}^2 = U_n + V_n \xrightarrow[n \to \infty]{P} E[X_1^2] - \mu^2 = \sigma^2.$$

Thus, the sample variance, S_n^2, is a consistent estimator of σ^2.

Think About It What about the other version of the sample variance, namely,

$$S_{n-1}^2 = \frac{1}{n-1} \sum_{j=1}^{n} (X_j - \overline{X})^2 = \frac{n}{n-1} S_n^2?$$

Taking $U_n = \frac{n}{n-1}$ and $V_n = S_n^2$, it follows from part (b) in Theorem 8.4 that

$$U_n V_n = S_{n-1}^2 \xrightarrow[n \to \infty]{P} (1)\sigma^2 = \sigma^2,$$

so that (not surprisingly) S_{n-1}^2 is also a consistent estimator of σ^2.

Example 8.7 Consistency of C.D.F. Estimator for Normal Distribution

Let $X_1, \ldots, X_n$ be a random sample from a $n\,(\mu, \sigma^2)$ distribution, and let t be arbitrary. One natural estimator for the c.d.f. of this $n\,(\mu, \sigma^2)$ distribution is given by $\widehat{F}_X(t) = \widehat{P}(X \leq t) = \Phi\left(\frac{t-\overline{X}}{S_n}\right)$, where $\overline{X}$ and S_n^2 are the sample mean and sample variance (division by n), respectively, and $\Phi\,(\cdot)$ is the c.d.f. for the standard normal distribution. Then we know the following:

(i) $\overline{X} \xrightarrow[n\to\infty]{P} \mu$ (WLLN)

(ii) $S_n^2 \xrightarrow[n\to\infty]{P} \sigma^2$ (Example 8.6, since $E[X^4] < \infty$)

(iii) $V_n = S_n \xrightarrow[n\to\infty]{P} \sigma$ (continuous function $g(c) = \sqrt{c}$ in Theorem 8.3)

(iv) $U_n = t - \overline{X} \xrightarrow[n\to\infty]{P} t - \mu$ (continuous function $g(c) = t - c$ in Theorem 8.3)

(v) $\frac{t-\overline{X}}{S_n} = \frac{U_n}{V_n} \xrightarrow[n\to\infty]{P} \frac{t-\mu}{\sigma}$ (part (c) of Theorem 8.4, since $\sigma \neq 0$)

(vi) $\widehat{F}_X(t) = \Phi\left(\frac{t-\overline{X}}{S_n}\right) \xrightarrow[n\to\infty]{P} \Phi\left(\frac{t-\mu}{\sigma}\right) = F_X(t)$ for all t

(Theorem 8.3, since $g(c) = \Phi(c)$ is a continuous function)

Thus, $\widehat{F}_X(t)$ is a consistent estimator for the c.d.f. $F_X(t)$ for all t.

8.2 Convergence in Distribution

Let Q_n be a random variable depending on the integer n (usually related to the sample size for a statistical sample). In order to compute probabilities such as

$$P(a < Q_n \leq b) = G_n(b) - G_n(a), \quad \text{with } a < b,$$

we need to obtain the explicit form of the probability distribution for Q_n (or its c.d.f. $G_n(q)$). This can be very difficult in many settings, so that it is important to have ways to approximate these cumulative probabilities $G_n(q)$ when n is large (i.e., large sample size). This leads us to the concept of convergence in distribution.

Definition 8.3 Let $Q_1, Q_2, \ldots, Q_n, \ldots$ be a sequence of random variables with c.d.f.'s $G_1, G_2, \ldots, G_n, \ldots$, respectively, and let Q be a random variable with c.d.f. G (not depending on n). We say that Q_n **converges in distribution to Q as $n \to \infty$** if $\lim_{n\to\infty} G_n(x) = G(x)$ at every point x at which $G(x)$ is continuous.

We denote this convergence in distribution property by $Q_n \xrightarrow[n\to\infty]{d} Q$ and say that Q_n has an **asymptotic (limiting) distribution $(n \to \infty)$ with c.d.f. G**.

Thus, if $Q_n \xrightarrow[n\to\infty]{d} Q$, we can approximate (for large n) probabilities for Q_n by

$$P(a < Q_n \leq b) = G_n(b) - G_n(a) \approx G(b) - G(a) = P(a < Q \leq b), \quad \text{with } a < b.$$
$$(8.1)$$

Mathematical Moment 8   Convergence of a Useful Sequence

Theorem 8.5 Let a and b be arbitrary constants, not depending on n. Then

$$\lim_{n\to\infty} \left[1 + \frac{a}{n}\right]^{bn} = e^{ab}.$$

Proof Note that

$$\ln\left\{ \left[1 + \frac{a}{n}\right]^{bn} \right\} = bn\ln\left(1 + \frac{a}{n}\right) = \frac{\ln(1 + \frac{a}{n})}{\{1/bn\}}.$$

Thus

$$\lim_{n\to\infty} \ln\left\{ \left[1 + \frac{a}{n}\right]^{bn} \right\} = \frac{\lim_{n\to\infty}\left\{ \ln\left[1 + \frac{a}{n}\right] \right\}}{\lim_{n\to\infty}\{1/bn\}}.$$
$$(8.2)$$

Now, we know that

$$\lim_{n\to\infty}\left\{ \ln\left[1 + \frac{a}{n}\right] \right\} = \lim_{n\to\infty}\{1/bn\} = 0.$$

Thus, we can apply L'Hospital's Rule (Mathematical Moment 4) and take derivatives of both the numerator and denominator in (8.2) to obtain

$$\lim_{n\to\infty} \ln\left\{ \left[1 + \frac{a}{n}\right]^{bn} \right\} = \frac{\lim_{n\to\infty}\left[\frac{-\frac{a}{n^2}}{1 + \frac{a}{n}} \right]}{\lim_{n\to\infty}\left[-\frac{1}{bn^2} \right]} = \lim_{n\to\infty}\left[\frac{ab}{1 + \frac{a}{n}} \right] = ab.$$

It follows that $\lim_{n\to\infty}\left[1 + \frac{a}{n}\right]^{bn} = \lim_{n\to\infty} e^{\ln\left\{ \left[1 + \frac{a}{n}\right]^{bn} \right\}} = e^{ab}.$ ∎

Example 8.8 Maximum Order Statistic from a Uniform Distribution
Let $X_{(1)} \leq \ldots \leq X_{(n)}$ be the order statistics for a random sample of size n from the *Unif* $(0, \theta)$ distribution, with $\theta > 0$. The underlying p.d.f. is given by

$$f(x; \theta) = \frac{1}{\theta} I_{(0,\theta)}(x).$$

Let $Q_n = n(\theta - X_{(n)})$. For any $t \leq 0$, $P(Q_n \leq t) = G_n(t) = 0$, since $P(X_{(n)} \geq \theta) = 0$. For any $t > 0$, the c.d.f. for Q_n is given by

$$G_n(t) = P(Q_n \leq t) = P(n(\theta - X_{(n)}) \leq t)$$

$$= P\left(X_{(n)} \geq \theta - \frac{t}{n}\right) = 1 - P\left(X_{(n)} \leq \theta - \frac{t}{n}\right)$$

$$= 1 - \left\{P\left(X_1 \leq \theta - \frac{t}{n}\right)\right\}^n = 1 - \left\{\int_0^{\theta - \frac{t}{n}} \frac{1}{\theta} dx\right\}^n$$

$$= 1 - \left\{\frac{\theta - \frac{t}{n}}{\theta}\right\}^n = 1 - \left\{1 + \frac{\left(-\frac{t}{\theta}\right)}{n}\right\}^n.$$

From Mathematical Moment 8 with $a = -t/\theta$ and $b = 1$, it follows that

$$\lim_{n \to \infty} G_n(t) = \lim_{n \to \infty} \left[1 - \left\{1 + \frac{\left(-\frac{t}{\theta}\right)}{n}\right\}^n\right] = 1 - e^{-\frac{t}{\theta}}, \quad 0 < t < \infty.$$

Now, define the random variable Q to have c.d.f.

$$G(t) = 0, \qquad t \leq 0$$

$$= 1 - e^{-\frac{t}{\theta}}, \quad 0 < t < \infty.$$

Since $\lim_{n \to \infty} G_n(t) = G(t)$ for all t, it follows that $Q_n \overset{d}{\to} Q$ with c.d.f. G, as $n \to \infty$. To obtain the asymptotic $(n \to \infty)$ p.d.f. for Q_n, we can simply differentiate the asymptotic c.d.f. (since the variables are continuous) to obtain

$$g(t) = \frac{1}{\theta} e^{-\frac{t}{\theta}} I_{(0,\infty)}(t),$$

which we recognize to be the p.d.f. for the *Gamma* $(\alpha = 1, \beta = \theta)$ distribution. Another way to denote this fact is to write $n(\theta - X_{(n)}) \overset{d}{\to}$ *Gamma* $(\alpha = 1, \beta = \theta)$ distribution as $n \to \infty$; that is, the asymptotic (limiting) distribution $(n \to \infty)$ for $n(\theta - X_{(n)})$ is *Gamma* $(\alpha = 1, \beta = \theta)$.

8.2.1 Convergence of Moment Generating Functions

Remember that a probability distribution is uniquely determined by both its c.d.f. F and its moment generating function $M(t)$. This relationship carries over to asymptotic distributions as well, as the following theorem (stated without proof) demonstrates.

Theorem 8.6 Continuity Theorem Let $\{G_n\}$ be a sequence of c.d.f.'s with the corresponding sequence of moment generating functions $\{M_n\}$. Let G be a c.d.f. with associated moment generating function M. If $\lim_{n\to\infty} M_n(t) = M(t)$ for all t in an open interval containing 0, then $\lim_{n\to\infty} G_n(x) = G(x)$ at all points of continuity for $G(x)$.

Thus, if $\{W_n\}$ is a sequence of random variables with c.d.f.'s $\{G_n\}$ and moment generating functions $\{M_n\}$ such that $\lim_{n\to\infty} M_n(t) = M(t)$ for all t in an open interval containing 0, then $\lim_{n\to\infty} G_n(x) = G(x)$, where G is the c.d.f. associated with the moment generating function M. In other words, $W_n \xrightarrow[n\to\infty]{d} W$, where W is a random variable with c.d.f. G and moment generating function M; that is, the distribution of W is the limiting asymptotic $(n \to \infty)$ distribution of W_n.

This Continuity Theorem can be used to establish the most important approximation theorem in statistics, namely, the **Central Limit Theorem (CLT)**. First, we prove a special case of the CLT that establishes the fact that we can use the standard normal distribution to approximate Poisson probabilities.

Theorem 8.7 Normal Approximation to the Poisson Distribution Let $\{W_n\}$ be a sequence of Poisson random variables with parameters $\{\lambda_n\}$, where $0 \le \lambda_1 \le \lambda_2 \le \cdots$ such that $\lim_{n\to\infty} \lambda_n = \infty$. Then,

$$Z_n = \frac{W_n - E[W_n]}{\sqrt{Var(W_n)}} = \frac{W_n - \lambda_n}{\sqrt{\lambda_n}} \xrightarrow[n\to\infty]{d} Z,$$

where Z has a $n\,(0, 1)$ distribution.

Proof The moment generating function for the Poisson variable W_n is

$$M_{W_n}(t) = e^{\lambda_n(e^t - 1)} I_{(-\infty,\infty)}(t).$$

It follows that the moment generating function for Z_n is given by

$$M_{Z_n}(t) = E\left[e^{t Z_n}\right] = E\left[e^{t\left(\frac{W_n - \lambda_n}{\sqrt{\lambda_n}}\right)}\right] = e^{-t\frac{\lambda_n}{\sqrt{\lambda_n}}} E\left[e^{t\frac{W_n}{\sqrt{\lambda_n}}}\right]$$

$$= e^{-t\sqrt{\lambda_n}} M_{W_n}\left(\frac{t}{\sqrt{\lambda_n}}\right) = e^{-t\sqrt{\lambda_n}} e^{\lambda_n\left(e^{\frac{t}{\sqrt{\lambda_n}}} - 1\right)}.$$

Taking natural logarithms, we have

$$\ln(M_{Z_n}(t)) = -t\sqrt{\lambda_n} + \left[\lambda_n\left(e^{\frac{t}{\sqrt{\lambda_n}}} - 1\right)\right].$$

Replacing $e^{\frac{t}{\sqrt{\lambda_n}}}$ by its Taylor series expansion

$$1 + \frac{t}{\sqrt{\lambda_n}} + \frac{t^2}{2!\lambda_n} + \frac{t^3}{3!\lambda_n^{3/2}} + \frac{t^4}{4!\lambda_n^2} + \cdots \quad \left(\begin{array}{c} \text{the remaining terms are} \\ \text{of order } \lambda_n^{-5/2} \text{ or higher} \end{array} \right),$$

we have

$$\ln\left(M_{Z_n}(t)\right) = -t\sqrt{\lambda_n} + \left[t\sqrt{\lambda_n} + \frac{t^2}{2} + \frac{t^3}{6\sqrt{\lambda_n}} + \frac{t^4}{4!\lambda_n} \right] + \left(\begin{array}{c} \text{terms of order} \\ \lambda_n^{-3/2} \text{ or higher} \end{array} \right)$$

$$= \frac{t^2}{2} + \frac{t^3}{6\sqrt{\lambda_n}} + \frac{t^4}{4!\lambda_n} + \left(\text{terms of order } \lambda_n^{-3/2} \text{ or higher} \right).$$

Thus,

$$\lim_{n \to \infty} \ln M_{Z_n}(t) = \lim_{n \to \infty} \left[\frac{t^2}{2} + \frac{t^3}{6\sqrt{\lambda_n}} + \text{terms of order } \lambda_n^{-1} \text{ or higher} \right] = \frac{t^2}{2},$$

and

$$\lim_{n \to \infty} M_{Z_n}(t) = \lim_{n \to \infty} e^{\ln M_{Z_n}(t)} \overset{\text{continuous function}}{=} e^{\lim_{n \to \infty} \ln M_{Z_n}(t)} = e^{\frac{t^2}{2}}, \quad -\infty < t < \infty.$$

Since this limit is the moment generating function for a standard normal distribution, we have (via the Continuity Theorem 8.6) that

$$Z_n = \frac{W_n - \lambda_n}{\sqrt{\lambda_n}} \overset{d}{\to} Z \sim n(0, 1), \quad \text{as } n \to \infty. \ \blacksquare$$

The implication of Theorem 8.7 is that if $X \sim Poisson$ (λ), then the distribution of $W = \frac{X - \lambda}{\sqrt{\lambda}}$ can be approximated by the standard normal distribution for large values of λ.

Example 8.9 Normal Approximation to the Poisson Distribution

Suppose the probability of a blemish in a foot of produced wire filament is .001. If we purchase 9000 feet of this wire filament, what is the probability that we will have at least 12 blemishes in our purchase?

Let X be the number of blemishes in our purchased wire filament. The Poisson distribution is frequently used as an approximate probability model for such a setting, where the mean for the Poisson is taken to be $\lambda = 9000(.001) = 9$ (i.e., the expected number of blemishes in our purchased 9000 feet of wire filament when the probability of a single blemish in 1 foot of the wire filament is .001). That is, we take our probability model to be $X \sim Poisson$ $(\lambda = 9)$. Under this assumption, the probability of interest can be found by using the **R** function *ppois( )* with arguments *lambda* $= 9$ and *lower.tail* $= FALSE$. As this function will provide $P(X > q)$, we specify the argument $q = 11$ as well to get $P(X > 11) = P(X \geq 12)$ as follows:

> ppois($q = 11$, *lambda* = 9, *lower.tail* = FALSE)

[1] 0.1969916

$$\Rightarrow P(X \geq 12) = \sum_{x=12}^{\infty} \frac{9^x e^{-9}}{x!} = 1 - \sum_{x=0}^{11} \frac{9^x e^{-9}}{x!} = 1 - .8030 = .1970.$$

On the other hand, we can use Theorem 8.7 and the normal approximation to the Poisson distribution to obtain

$$P(X \geq 12) = P(\frac{X-9}{\sqrt{9}} \geq \frac{12-9}{3}) \overset{Z \sim n(0,1)}{\approx} P(Z \geq 1)$$
$$= 1 - P(Z \leq 1) = 1 - \Phi(1) = 1 - .8413 = .1587,$$

where $\Phi(1) = .8413$ is found using the **R** function *pnorm( )* with the argument $q = 1$ and *lower.tail = FALSE* as follows:

> pnorm(q = 1, lower.tail = FALSE)

[1] 0.1586553

We note that this normal approximation (.1587) is not a very accurate estimate of the true probability .1970. This is largely due to the fact that we are approximating a discrete distribution (Poisson) by the continuous normal distribution. In such cases, it is better to use the continuity correction, corresponding to treating the exact probability that $X = 12$ as spread out over the interval 11.5–12.5 for the normal approximation.

Using this continuity correction (see Fig. 8.1), we have

$$P(X \geq 12) = P(X \geq 11.5) = P\left(\frac{X-9}{\sqrt{9}} \geq \frac{11.5-9}{3}\right) \overset{Z \sim n(0,1)}{\approx} P(Z \geq .833)$$
$$= 1 - P(Z \leq .833) = 1 - \Phi(.833) = 1 - .7967 = .2033,$$

which is a much better approximation to the exact probability .1970.

8.2.2 Central Limit Theorem (CLT)

In Theorem 8.7 we saw that the normal distribution could be used to approximate probabilities for the Poisson distribution. However, the normal distribution plays a much more important role in approximating probabilities for sums of random

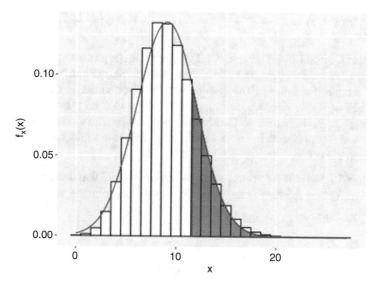

Fig. 8.1 Using the normal probability density function with continuity correction to approximate $P(X \geq 12)$ for $X \sim Poisson$ $(\lambda = 9)$

variables than just this special case of the Poisson distribution. In fact, we have a quite remarkable result that the normal distribution can be used to approximate probabilities for *any* sum of independent and identically distributed random variables (i.e., a random sample) with finite variance. We state this result without proof, only noting that it can be established by taking the same approach we used in our proof of Theorem 8.7.

Theorem 8.8 Central Limit Theorem (CLT) Let $X_1, \ldots, X_n$ be a sequence of independent and identically distributed random variables (i.e., random sample) with mean μ and variance $\sigma^2 < \infty$. Let $S_n = \sum_{i=1}^{n} X_i$ and $\overline{X} = S_n/n$ (the sample mean). Then,

$$\frac{S_n - E[S_n]}{\sqrt{Var(S_n)}} = \frac{S_n - n\mu}{\sigma\sqrt{n}} = \frac{\sqrt{n}(\overline{X} - \mu)}{\sigma} \xrightarrow[n\to\infty]{d} Z \sim n(0,1).$$

Thus, for any t, we have

$$P(S_n \leq t) = P\left(\frac{S_n - n\mu}{\sigma\sqrt{n}} \leq \frac{t - n\mu}{\sigma\sqrt{n}}\right) \approx \Phi\left(\frac{t - n\mu}{\sigma\sqrt{n}}\right) \qquad (8.3)$$

or, equivalently,

$$P(\bar{X} \le t) = P\left(\frac{\sqrt{n}(\bar{X} - \mu)}{\sigma} \le \frac{\sqrt{n}(t - \mu)}{\sigma}\right) \approx \Phi\left(\frac{\sqrt{n}(t - \mu)}{\sigma}\right). \qquad (8.4)$$

We use this Central Limit Theorem (CLT) to approximate probabilities for sums and averages of n independent and identically distributed random variables (i.e., a random sample) for finite n. The goodness of the approximation depends on how large n is, but it also depends on properties of the underlying distribution for the sample $X_1, \ldots, X_n$. In particular, the accuracy of the approximation depends on both the symmetry of the underlying X distribution and how rapidly the tails of the distribution die off.

Example 8.10 Normal Approximation to the Binomial Distribution

Let $X_1, \ldots, X_n$ be a random sample from a Bernoulli distribution with parameter p. Then $\sum_{i=1}^{n} X_i \sim Binom\,(n, p)$ and

$$P(a \le \sum_{i=1}^{n} X_i \le b) = \sum_{x=a}^{b} \binom{n}{x} p^x (1 - p)^{n-x}$$

$$= P\left(\frac{a - np}{\sqrt{np(1 - p)}} \le \frac{\sum_{i=1}^{n} X_i - np}{\sqrt{np(1 - p)}} \le \frac{b - np}{\sqrt{np(1 - p)}}\right)$$

$$\stackrel{CLT}{\approx} \Phi\left(\frac{b - np}{\sqrt{np(1 - p)}}\right) - \Phi\left(\frac{a - np}{\sqrt{np(1 - p)}}\right),$$

for any integers a and b such that $0 \le a < b \le n$. This approximation is best if p is not near 0 or 1 and is generally considered acceptable if $np > 5$ and $n(1 - p) > 5$.

As a specific example, suppose a fair six-sided die is rolled 100 times. We want to find the probability that the face showing a six turns up between 15 and 20 times, inclusive. Let X be the number of sixes in the 100 rolls. Then $X \sim Binom\,(100, \frac{1}{6})$, so that the exact probability of interest is

$$P(15 \le X \le 20) = \sum_{x=15}^{20} \binom{100}{x} \left(\frac{1}{6}\right)^x \left(\frac{5}{6}\right)^{100-x} = P(X \le 20) - P(X \le 14)$$

$$= .8481 - .2874 = .5607,$$

where these exact probabilities are obtained by using the difference between two calls to the **R** function *pbinom( )* with arguments *size = 100* and *prob = 1/6* as follows:

> pbinom(q = 20, size = 100, prob = 1/6) – pbinom(q = 14, size = 100, prob = 1/6)

[1] 0.5606912

How well does the normal approximation do in this setting (p is not very close to 0 or 1 and $n = 100$)? Using the approximation, we have

$$P(15 \leq X \leq 20) \overset{CLT}{\approx} \Phi\left(\frac{20 - 100(1/6)}{\sqrt{100(1/6)(5/6)}}\right) - \Phi\left(\frac{15 - 100(1/6)}{\sqrt{100(1/6)(5/6)}}\right)$$
$$= \Phi(.89) - \Phi(-.45) = .8133 - .3264 = .4869.$$

While this approximation is reasonable, it is not as close to the true value as we might like. Once again, however, using the continuity correction for approximating the discrete binomial distribution with the continuous normal distribution dramatically improves the approximation. Employing the continuity correction, we see that

$$P(15 \leq X \leq 20) = P(14.5 \leq X \leq 20.5)$$
$$\overset{CLT}{\approx} \Phi\left(\frac{20.5 - 100(1/6)}{\sqrt{100(1/6)(5/6}}\right) - \Phi\left(\frac{14.5 - 100(1/6)}{\sqrt{100(1/6)(5/6)}}\right)$$
$$= \Phi(1.03) - \Phi(.58) = .8485 - .2810 = .5675,$$

which is, indeed, an excellent approximation for the exact probability .5607.

Example 8.11 Properties of the Sample Mean, $\overline{X}$

Let $X_1, \ldots, X_n$ be a random sample from an arbitrary probability distribution with mean μ and variance $\sigma^2 < \infty$. The sample mean $\overline{X} = \frac{1}{n}\sum_{i=1}^{n} X_i$ is often used as an estimator for the population mean μ. What properties have we learned about this statistic $\overline{X}$ that do not depend on the specific form of the underlying X distribution?

(i) $\overline{X} \overset{P}{\underset{n\to\infty}{\to}} \mu$ (WLLN)

(ii) $P\left(|\overline{X} - \mu| > t\right) \leq \frac{\sigma^2}{nt^2}$, $\forall t > 0$ (Chebyshev's Inequality)

(iii) $\frac{\sqrt{n}(\overline{X}-\mu)}{\sigma} \overset{d}{\underset{n\to\infty}{\to}} Z \sim n(0,1)$ (CLT).

Although Chebyshev's Inequality provides us with an upper bound about how accurate $\overline{X}$ is as an estimator of μ, the CLT is more informative. We use the CLT to approximate this accuracy as follows. Let $t > 0$ be arbitrary. Then, we have

$$\gamma = P\left(|\overline{X} - \mu| < t\right) = P\left(-t < \overline{X} - \mu < t\right) = P\left(-\frac{\sqrt{n}t}{\sigma} < \frac{\sqrt{n}(\overline{X} - \mu)}{\sigma} < \frac{\sqrt{n}t}{\sigma}\right)$$
$$\overset{CLT}{\approx} \Phi\left(\frac{\sqrt{n}t}{\sigma}\right) - \Phi\left(-\frac{\sqrt{n}t}{\sigma}\right) \overset{symmetry}{=} 2\Phi\left(\frac{\sqrt{n}t}{\sigma}\right) - 1.$$

$$(8.5)$$

For given t, n, and σ^2, we can use (8.5) to find γ. For example, if $n = 25$ and $\sigma^2 = 4$, then the probability that $\overline{X}$ deviates from the unknown population mean μ by less than 1 is approximately

$$\gamma = P\left(|\overline{X} - \mu| < 1\right) \approx 2\Phi\left(\frac{\sqrt{25}(1)}{\sqrt{4}}\right) - 1 = 2\Phi(2.5) - 1 = 2(.9938) - 1$$

$$= .9976.$$

This is a much more informative statement about the accuracy of $\overline{X}$ than is provided by the Chebyshev's lower bound assertion that

$$P\left(|\overline{X} - \mu| < 1\right) \geq 1 - \frac{4}{25(1)^2} = 1 - .16 = .84.$$

In addition, when σ^2 is known, we can use the CLT to provide an approximate idea as to how large a sample size n is needed in order to obtain a prescribed accuracy for $\overline{X}$ as an estimator of μ. Suppose $\sigma^2 = 25$. Approximately how large must n be so that we are 95% confident that $\overline{X}$ will be within 1 of μ? Using the CLT, we have that

$$P\left(|\overline{X} - \mu| < 1\right) \approx 2\Phi\left(\frac{\sqrt{n}(1)}{\sqrt{25}}\right) - 1 = .95$$

$$\Rightarrow 2\Phi\left(\frac{\sqrt{n}}{5}\right) = 1.95 \Rightarrow \Phi\left(\frac{\sqrt{n}}{5}\right) = .975.$$

We can then use the **R** function $qnorm()$ to find $z_{0.975}$ by specifying the p argument as follows:

```
> qnorm(p = 0.975)
```

[1] 1.959964

$$\Rightarrow \frac{\sqrt{n}}{5} \approx z_{.975} = 1.96 \Rightarrow \sqrt{n} \approx 5(1.96) = 9.8 \Rightarrow n \approx (9.8)^2 = 96.04.$$

Thus, if $\sigma^2 = 25$, a sample size of $n = 96$ or 97 will be sufficient to have approximate probability that $\overline{X}$ deviates from the unknown population mean μ by no more than 1, regardless of the form of the underlying X distribution.

We now turn our attention to two very important results that extend the usefulness of the Central Limit Theorem to many additional settings.

8.2.3   Slutsky's Theorem

Convergence in probability properties can at times be used effectively to show the equivalence of asymptotic distributions for appropriately related random variables. One such situation is described in the following result known as Slutsky's Theorem.

Theorem 8.9 Slutsky's Theorem Suppose that $U_n \xrightarrow[n \to \infty]{d} U$, $V_n \xrightarrow[n \to \infty]{P} c$, and $W_n \xrightarrow[n \to \infty]{P} d$, for constants c and d. Then $V_n U_n + W_n \xrightarrow[n \to \infty]{d} cU + d$.

Proof See, for example, Section 20.6 in Cramér (1946).

Corollary 8.1 Suppose $U_n \xrightarrow[n \to \infty]{d} U$ and $T_n - U_n \xrightarrow[n \to \infty]{P} 0$. Then, $T_n \xrightarrow[n \to \infty]{d} U$, as well. That is, U_n and T_n have the same asymptotic distribution as $n \to \infty$.

Proof Take $V_n = 1$ and $W_n = T_n - U_n$ in Slutsky's Theorem 8.9. ∎

Example 8.12 Asymptotic Normality of the Standardized Sample Mean
Let $X_1, \ldots, X_n$ be a random sample from a probability distribution with mean μ and variance $\sigma^2 < \infty$. Let $\overline{X} = \frac{1}{n} \sum_{i=1}^{n} X_i$ and $S_{n-1}^2 = \frac{1}{n-1} \sum_{i=1}^{n} (X_i - \overline{X})^2$ be the associated sample mean and sample variance (with divisor $n - 1$). We have previously established the following facts:

Fact 1: From the Central Limit Theorem

$$\frac{\sqrt{n}(\overline{X} - \mu)}{\sigma} \xrightarrow{d} Z \sim n(0, 1), \quad \text{as } n \to \infty.$$

Fact 2: We have shown (Example 8.6) that $S_{n-1}^2 \xrightarrow[n \to \infty]{P} \sigma^2$. From the various rules about convergence in probability stated in Theorem 8.4, it follows that

$$S_{n-1}^2 \xrightarrow[n \to \infty]{P} \sigma^2 \Rightarrow S_{n-1} \xrightarrow[n \to \infty]{P} \sigma \Rightarrow \frac{\sigma}{S_{n-1}} \xrightarrow[n \to \infty]{P} 1.$$

Taking $U_n = \frac{\sqrt{n}(\overline{X} - \mu)}{\sigma}$, $V_n = \frac{\sigma}{S_{n-1}}$, and $W_n = 0$, it follows immediately from Slutsky's Theorem 8.9 that

$$V_n U_n + W_n = \frac{\sigma}{S_{n-1}} \left(\frac{\sqrt{n}(\overline{X} - \mu)}{\sigma} \right) = \frac{\sqrt{n}(\overline{X} - \mu)}{S_{n-1}} \xrightarrow{d} Z \sim n(0, 1), \quad \text{as } n \to \infty.$$

Note When the underlying distribution is normal, it can be shown that the random variable considered in Example 8.12, namely,

$$Q_n = \frac{\sqrt{n}(\overline{X} - \mu)}{S_{n-1}},$$

has a well-known distribution called the t-distribution with $n - 1$ degrees of freedom for all finite values of the sample size n. The result in Example 8.12 then implies that the t-distribution with $n - 1$ degrees of freedom converges to the standard normal distribution as $n \to \infty$. Moreover, Example 8.12 establishes the more robust result that the asymptotic ($n \to \infty$) distribution for Q_n is standard normal even without this underlying normality requirement!

The second important extension of the Central Limit Theorem deals with functions of random variables that themselves have asymptotic normal distributions.

8.2.4   Delta Method

Theorem 8.10 Delta Method Let W_n be a random variable such that

$$\sqrt{n}(W_n - \theta) \xrightarrow{d} n\left(0, \gamma^2\right) \quad \text{as } n \to \infty, \tag{8.6}$$

and let $g(x)$ be a function for which $g'(x)$ exists and is continuous in some neighborhood of θ. Then,

$$\sqrt{n}[g(W_n) - g(\theta)] \xrightarrow{d} n\left(0, [g'(\theta)]^2 \gamma^2\right) \quad \text{as } n \to \infty,$$

provided $g'(\theta) \neq 0$.

Proof From the Mean Value Theorem, there exists a ξ_n between θ and W_n such that

$$g(W_n) - g(\theta) = (W_n - \theta)g'(\xi_n). \tag{8.7}$$

Now, it follows from Exercise 8.20 that the convergence in distribution assumption in (8.6) implies that $W_n \xrightarrow{P} \theta$ as $n \to \infty$, which, in turn, implies that $\xi_n \xrightarrow{P} \theta$ as $n \to \infty$. Since $g'(\cdot)$ is continuous at θ, it follows from Theorem 8.3 that

$$g'(\xi_n) \xrightarrow{P} g'(\theta) \quad \text{as } n \to \infty. \tag{8.8}$$

From (8.7), we have

$$\sqrt{n}[g(W_n) - g(\theta)] = \sqrt{n}[W_n - \theta]g'(\xi_n) = \{\sqrt{n}[W_n - \theta]g'(\theta)\}\left\{\frac{g'(\xi_n)}{g'(\theta)}\right\}. \tag{8.9}$$

It then follows from (8.8) and Slutsky's Theorem that $\sqrt{n}[g(W_n) - g(\theta)]$ has the same asymptotic $(n \to \infty)$ distribution as $\sqrt{n}[W_n - \theta]g'(\theta)$. But our assumption in (8.6) implies that $\sqrt{n}[W_n - \theta]\, g'(\theta) \xrightarrow[n \to \infty]{d} n\,(0, \gamma^2[g'(\theta)]^2)$. Hence, we have the desired result that

$$\sqrt{n}[g(W_n) - g(\theta)] \xrightarrow[n \to \infty]{d} n\,(0, \gamma^2[g'(\theta)]^2). \quad \blacksquare$$

Thus, for W_n and $g(\cdot)$ satisfying the conditions of Theorem 8.10, we can use the $n\left(g(\theta), \frac{\gamma^2[g'(\theta)]^2}{n}\right)$ distribution to approximate probabilities for the distribution of $g(W_n)$ when n is large.

Example 8.13 Asymptotic Normality for Functions of the Sample Mean $\overline{X}$
Let $X_1, \ldots, X_n$ be a random sample from a distribution with mean μ and finite variance σ^2. Then, by the Central Limit Theorem, we have that

$$\frac{\sqrt{n}(\overline{X} - \mu)}{\sigma} \xrightarrow{d} n\,(0, 1), \quad \text{as } n \to \infty,$$

which implies that

$$\sqrt{n}(\overline{X} - \mu) \xrightarrow{d} n\,(0, \sigma^2), \quad \text{as } n \to \infty.$$

Letting $W_n = \overline{X}$, $\theta = \mu$, and $\gamma^2 = \sigma^2$, it follows from the Delta Method (Theorem 8.10) that

$$\sqrt{n}[g(\overline{X}) - g(\mu)] \xrightarrow{d} n\,(0, [g'(\mu)]^2\sigma^2) \quad \text{as } n \to \infty,$$

for any $g(x)$ such that $g'(x)$ is continuous and non-zero at $x = \mu$. This means that we can use the $n(g(\mu), \frac{\sigma^2[g'(\mu)]^2}{n})$ distribution to approximate probabilities for the distribution of $g(\overline{X})$ when n is large. For example, if we take $g(x) = \frac{1}{x}$, then $g'(x) = -\frac{1}{x^2}$ exists and is continuous for all $x \neq 0$, which implies that we can use the $n\left(\frac{1}{\mu}, \frac{\sigma^2}{n\mu^4}\right)$ distribution to approximate probabilities for $g(\overline{X}) = \frac{1}{\overline{X}}$ when n is large, provided $\mu \neq 0$.

Example 8.14 Let $X_1, \ldots, X_n$ be a random sample from the continuous distribution with p.d.f.

$$f_X(x|\theta) = \theta x^{\theta-1} I_{(0,1)}(x) I_{(0,\infty)}(\theta).$$

In this setting, one of the standard estimators (the *maximum likelihood estimator*) for θ is $\widehat{\theta} = -\dfrac{n}{\sum\limits_{i=1}^{n} \ln X_i}$. We wish to explore how to approximate the probability distribution for $\widehat{\theta}$ when n is large.

In Example 6.7, we had shown that if $X \sim f_X(x|\theta)$, then $W = -2\theta \ln(X) \sim$ Gamma $(\alpha = 1, \beta = 2)$. Thus, we have $E[-\ln X] = \frac{1}{\theta}$ and $Var(-\ln X) = \frac{1}{\theta^2}$. Let $Y_i = -\ln X_i$, for $i = 1, \ldots, n$, and set

$$\overline{Y} = \frac{1}{n} \sum_{i=1}^{n} Y_i = -\frac{1}{n} \sum_{i=1}^{n} \ln X_i.$$

It follows from the Central Limit Theorem applied to the Y's that

$$\sqrt{n}\left(\overline{Y} - \frac{1}{\theta}\right) = \sqrt{n}\left(\frac{-\sum_{i=1}^{n} \ln X_i}{n} - \frac{1}{\theta}\right) \xrightarrow{d} n\left(0, \frac{1}{\theta^2}\right) \quad \text{as } n \to \infty.$$

Taking $g(x) = \frac{1}{x}$, it follows from Example 8.13 that

$$\sqrt{n}\left[g(\overline{Y}) - g\left(\frac{1}{\theta}\right)\right] = \sqrt{n}\left[-\frac{n}{\sum_{i=1}^{n} \ln X_i} - \theta\right] = \sqrt{n}\left(\widehat{\theta} - \theta\right)$$

$$\xrightarrow{d} n\left(0, \frac{1}{\theta^2}\left[g'\left(\frac{1}{\theta}\right)\right]^2\right) = n\left(0, \theta^2\right) \quad \text{as } n \to \infty.$$

Thus, we can approximate the probability distribution for $\widehat{\theta} = -\dfrac{n}{\sum\limits_{i=1}^{n} \ln X_i}$ by the $n\left(0, \frac{\theta^2}{n}\right)$ distribution when n is large.

8.3 Exercises

8.1. Let $\widehat{p}$ be the percentage of successes obtained in n independent Bernoulli trials, each with probability of success, p. Show that $\widehat{p} \xrightarrow{P} p$ as $n \to \infty$, so that $\widehat{p}$ is a consistent estimator for p.

8.2. Let $F_n(t) = \dfrac{\#X's \leq t}{n}$ be the empirical c.d.f. for a random sample $X_1, \ldots, X_n$ of size n from a probability distribution with c.d.f. $F_X(x)$. Show that $F_n(t) \xrightarrow{P} F_X(t)$, as $n \to \infty$, for every fixed t. Thus, the empirical c.d.f. is a consistent estimator for the probability $P(X \leq t) = F_X(t)$ for every fixed t.

8.3. Let $X_1, \ldots, X_n$ be a random sample of size n from a probability distribution with $\mu_t = E[X^t]$, $t = 1, 2, \ldots$ (provided they exist). If k is an integer such that $\mu_{2k} < \infty$, show that

$$\frac{1}{n} \sum_{i=1}^{n} X_i^k \xrightarrow{P} \mu_k = E[X^k] \quad \text{as } n \to \infty.$$

That is, if $\mu_{2k} < \infty$, the kth sample moment, $\frac{1}{n}\sum_{i=1}^{n} X_i^k$, converges in probability to (i.e., is a consistent estimator for) the kth population moment, $E[X^k]$, as $n \to \infty$.

8.4. Let $X_1, \ldots, X_n$ be a random sample from the continuous distribution with p.d.f.

$$f_X(x|\theta) = \theta x^{\theta-1} I_{(0,1)}(x) I_{(0,\infty)}(\theta).$$

Consider the estimator $\widehat{\theta}_2 = -\frac{n}{\sum_{i=1}^{n} \ln X_i}$. ($\widehat{\theta}_2$ is called the *maximum likelihood estima-tor* for θ.)

(a) Show that $E[-\ln X_1] = \frac{1}{\theta}$.

(b) Let $Y_i = -\ln X_i$, for $i = 1, \ldots, n$. Prove that $-\frac{\sum_{j=1}^{n} \ln X_j}{n} \xrightarrow[n\to\infty]{P} \frac{1}{\theta}$.

(c) Show that $\widehat{\theta}_2 \xrightarrow[n\to\infty]{P} \theta$. Hence, $\widehat{\theta}_2$ is a consistent estimator for θ.

(d) From Example 8.5 and part (c) of this exercise, we see that both $\widehat{\theta}_1 = \frac{\overline{X}}{1-\overline{X}}$ and $\widehat{\theta}_2 = -\frac{n}{\sum_{i=1}^{n} \ln X_i}$ are consistent estimators for θ. Discuss what criteria you might use to help determine which of the two estimators is preferred.

8.5. Let $X_1, \ldots, X_n$ be a random sample from the *Bernoulli* (θ) distribution with $0 \le \theta \le 1$.

(a) Argue that $\widehat{\theta} = \overline{X} = \frac{1}{n}\sum_{j=1}^{n} X_j$ is a consistent estimator for θ (i.e., $\widehat{\theta} \xrightarrow[n\to\infty]{P} \theta$).

(b) Suggest a natural estimator for $\sigma^2 = Var(X_1)$, and show that it is a consistent estimator.

8.6. Prove Theorem 8.4.

8.7. Consider the same setting as in Example 8.7. Another natural estimator for the c.d.f. $F_X(t)$ is given by

$$\widehat{F^*}_X(t) = \widehat{P}(X \le t) = \Phi\left(\frac{\sqrt{n-1}}{\sqrt{n}}\frac{t-\overline{X}}{S_n}\right) = \Phi\left(\frac{t-\overline{X}}{S_{n-1}}\right),$$

where S_{n-1}^2 is the form of the sample variance with division by $n-1$ rather than n. Argue that $\widehat{F^*}_X(t)$ is also a consistent estimator for the c.d.f. $F_X(t)$ for all t.

8.8. Let $X_1, \ldots, X_n$ be a random sample from a *Gamma* (α, β) distribution with $\alpha > 0$ and $\beta > 0$. One pair of estimators (*method of moments estimators*) for α and β are given by

$$\widehat{\alpha} = \frac{\overline{X}^2}{S_n^2} \quad \text{and} \quad \widehat{\beta} = \frac{S_n^2}{\overline{X}},$$

respectively, where $\overline{X}$ and S_n^2 are the sample mean and sample variance (division by n) for the random sample. Show that $\widehat{\alpha}$ and $\widehat{\beta}$ are consistent estimators for α and β, respectively. (Why are $\widehat{\alpha}$ and $\widehat{\beta}$, in some sense, "natural" estimators for α and β, respectively?)

8.9. Let $X_1, \ldots, X_n$ be a random sample from the continuous distribution with p.d.f.

$$f_X(x; p_1, p_2) = p_1 e^{-x} + p_2 x e^{-x} + .5(1 - p_1 - p_2) e^{-x/2} I_{(0,\infty)}(x),$$

where p_1 and p_2 are unknown parameters satisfying $0 \le p_1 \le 1$, $0 \le p_2 \le 1$, and $(p_1 + p_2) \le 1$.

(a) One pair of estimators (*method of moments (MOM) estimators*) for p_1 and p_2 are

$$\widetilde{p}_1 = 2 - \overline{X} \quad \text{and} \quad \widetilde{p}_2 = 3\overline{X} - 2 - \frac{1}{2n} \sum_{j=1}^{n} X_j^2,$$

respectively. Show that $\widetilde{p}_1$ and $\widetilde{p}_2$ are consistent estimators for p_1 and p_2, respectively. (Why are $\widetilde{p}_1$ and $\widetilde{p}_2$, in some sense, "natural" estimators for p_1 and p_2, respectively?)

(b) Discuss why these *MOM* estimators are not reasonable for this setting, in spite of the fact that they are consistent.

8.10. Let $X_1, \ldots, X_n$ be a random sample from the Poisson distribution with parameter $\theta > 0$. Since $E[X] = \theta$, it is natural to estimate θ by $\widehat{\theta}(X_1, \ldots, X_n) = \overline{X}$, the sample mean.

(a) Find $P(X_1 \le 1)$ as a function of θ, say $h(\theta)$.

(b) What is a natural estimator of $h(\theta)$, say $\widehat{h}(\theta)$?

(c) Show that the estimator $\widehat{h}(\theta)$ in part (b) is a consistent estimator for $h(\theta)$.

8.11. Let $X_{(1)} \le \ldots \le X_{(n)}$ be the order statistics for a random sample of size n from the logistic distribution with p.d.f.

$$f_X(x) = \frac{e^{-x}}{(1 + e^{-x})^2} I_{(-\infty,\infty)}(x).$$

(a) Find the distribution function for $X_{(n)}$.

(b) Find the limiting asymptotic $(n \to \infty)$ distribution for $W_n = X_{(n)} - \ln(n)$.

(c) Let W have the limiting asymptotic distribution obtained in part (b). Find the distribution of $V = e^{-W}$.

(d) Obtain the moment generating function for W defined in part (c).

8.12. A fair six-sided die is rolled 100 times. What is the approximate probability that the sum of the face values for the 100 rolls is less than 300?

Think About It Would it be easy to obtain the exact probability for this event? How would you approach that problem?

8.13. Let $X_1,\ldots,X_n$ be a random sample from the Bernoulli distribution with parameter p. Then the sample percentage of successes, $\hat{p} = \frac{1}{n}\sum_{i=1}^{n} X_i$, is the standard estimator for the parameter p. Show that the statistic

$$W_n = \frac{\sqrt{n}(\hat{p}-p)}{\sqrt{\hat{p}(1-\hat{p})}} \xrightarrow[n\to\infty]{d} Z \sim n\,(0,1).$$

8.14. Let $X_1,\ldots,X_n$ be a random sample from the continuous distribution with p.d.f.

$$f_X(x|\theta) = \theta x^{\theta-1} I_{(0,1)}(x) I_{(0,\infty)}(\theta),$$

and let $\hat{\theta} = -\dfrac{n}{\sum_{i=1}^{n}\ln X_i}$ be the *maximum likelihood estimator* for θ discussed in Exercise 8.4. Define the random variable

$$W_n = \frac{\sqrt{n}(\hat{\theta}-\theta)}{\hat{\theta}} = \sqrt{n} - \sqrt{n}\,\frac{\theta}{\hat{\theta}}.$$

Find the asymptotic ($n \to \infty$) distribution of W_n.

8.15. Let $X_1,\ldots,X_n$ be a random sample from the $n\,(\theta,\sigma^2)$ distribution, where σ^2 is known. We know that $\overline{X}$ is a natural estimator for θ. Let t_0 be arbitrary, but fixed, and let

$$g(\theta) = P(X \le t_0) = \Phi\!\left(\frac{t_0 - \theta}{\sigma}\right)$$

be the value of the c.d.f. for the $n\,(\theta,\sigma^2)$ distribution at t_0. A natural estimator for $g(\theta)$ is then

$$\hat{g}(\theta) = \hat{P}(X \le t_0) = g\!\left(\hat{\theta}\right) = g(\overline{X}) = \Phi\!\left(\frac{t_0 - \overline{X}}{\sigma}\right).$$

Find the form of the asymptotic ($n \to \infty$) distribution of

$$\sqrt{n}\left[g\left(\hat{\theta}\right) - g(\theta)\right] = \sqrt{n}\left[\Phi\left(\frac{t_0 - \overline{X}}{\sigma}\right) - \Phi\left(\frac{t_0 - \theta}{\sigma}\right)\right].$$

8.16. Let $X_1,\ldots,X_n$ be a random sample from a distribution with c.d.f. $F_X(x)$. The *empirical (or sample) c.d.f.* for these observations is defined by

$$F_n(x) = \frac{\#X's \leq x}{n} \quad \text{for all } x.$$

(a) For x fixed, find $E[F_n(x)]$ and $Var(F_n(x))$.

(b) For any fixed x, show that $F_n(x) \overset{P}{\to} F_X(x)$ as $n \to \infty$; that is, $F_n(x)$ is a consistent estimator of $F_X(x)$.

(c) For any fixed x, show that

$$\frac{F_n(x) - E[F_n(x)]}{\sqrt{Var(F_n(x))}} \overset{d}{\to} Z \sim n\,(0,1) \quad \text{as } n \to \infty.$$

(d) Argue that

$$\frac{\sqrt{n}[F_n(t) - F(t)]}{\sqrt{F_n(t)[1 - F_n(t)]}} \overset{d}{\to} Z \sim n\,(0,1) \quad \text{as } n \to \infty.$$

for any fixed t.

8.17. Let $X_1,\ldots,X_n$ be a random sample from a distribution with mean μ, variance σ^2, and $E[X^{2k}] < \infty$ for some integer $k \geq 2$. Define

$$V_n = \sum_{j=1}^{k}\sum_{i=1}^{n}\frac{X_i^j}{n^j}.$$

Show that $\frac{\sqrt{n}[V_n - \mu]}{\sigma} \overset{d}{\to} Z \sim n\,(0,1)$ as $n \to \infty$.

8.18. Suppose that $T_n \overset{d}{\underset{n\to\infty}{\to}} T$ and $E\left[(S_n - T_n)^2\right] \underset{n\to\infty}{\to} 0$. Show that $S_n \overset{d}{\underset{n\to\infty}{\to}} T$ as well.

8.19. Let $X_{(1)j} \leq X_{(2)j} \leq \cdots \leq X_{(n)j}$, for $j = 1, \ldots, k$, be the order statistics for k independent random samples from the probability distribution with p.d.f.

$$f_X(x) = 1I_{(0,1)}(x).$$

Let $V_{k,n} = \sum_{j=1}^{k} X_{(n)j}$. Show that we can approximate the c.d.f. for the probability distribution of $V_{k,n}$ by

$$P(V_{k,n} \leq t) \approx \Phi \left(\frac{t - \frac{kn}{n+1}}{\sqrt{\frac{kn}{(n+1)^2(n+2)}}} \right)$$

for large values of k.

8.20. Let W_n be a random variable such that

$$\sqrt{n}(W_n - \theta) \xrightarrow{d} n(0, \gamma^2) \quad \text{as } n \to \infty,$$

for some finite γ^2. Show that $W_n \xrightarrow{P} \theta$ as $n \to \infty$.

8.21. Let $X_{(1)} \leq \ldots \leq X_{(n)}$ be the order statistics for a random sample of size n from the continuous distribution with p.d.f.

$$f_X(x) = \frac{1}{2\theta} I_{(-\theta,\theta)}(x) I_{(0,\infty)}(\theta).$$

(a) Find the p.d.f. for $X_{(1)}$ and obtain $E[X_{(1)}]$ and $Var(X_{(1)})$.
(b) Without first obtaining the p.d.f. for $X_{(n)}$, find $E[X_{(n)}]$ and $Var(X_{(n)})$.
(c) Show that $\frac{X_{(1)}+X_{(n)}}{2} \xrightarrow{P} 0$ as $n \to \infty$.

8.22. Let $X_1, \ldots, X_n$ be a random sample of size n from the continuous distribution with p.d.f.

$$f_X(x) = \theta x^{\theta-1} I_{(0,1)}(x) I_{(0,\infty)}(\theta).$$

Let $W = \frac{\overline{X}}{1-\overline{X}}$, where $\overline{X} = \frac{1}{n} \sum_{i=1}^{n} X_i$ is the sample mean. Show that $W \xrightarrow{P} \theta$ as $n \to \infty$.

8.23. Let $X_1, \ldots, X_n$ be a random sample of size n from the continuous distribution with p.d.f.

$$f_X(x) = \alpha x^{\alpha-1} I_{(0,1)}(x) I_{(0,\infty)}(\alpha).$$

Let $W_n = \prod_{i=1}^{n} X_i^{\frac{1}{n}}$.

(a) Show that $W_n \xrightarrow{P} e^{-\frac{1}{\alpha}}$ as $n \to \infty$. [Hint: Consider the convergence of $\ln(W_n)$.]
(b) Specify a function of W_n that converges in probability to $E[X]$ as $n \to \infty$, and prove the result.

8.24. Let $X \sim n(\mu_1, \sigma^2)$ and let $Y \sim n(\mu_2, \sigma^2)$, with $-\infty < \mu_1 < \infty$, $-\infty < \mu_2 < \infty$, and $0 < \sigma^2 < \infty$.

(a) What is the probability distribution of $V = Y - X$? Justify your answer.

(b) Show that $P(X < Y) = \Phi\left(\frac{\mu_2 - \mu_1}{\sqrt{2}\,\sigma}\right)$.

(c) Let $X_1, \ldots, X_m$ and $Y_1, \ldots, Y_n$ be independent random samples of sizes m and n from $n\,(\mu_1, \sigma^2)$ and $n\,(\mu_2, \sigma^2)$, respectively. Define

$$\overline{X} = \sum_{i=1}^{m} X_i/m, \quad \overline{Y} = \sum_{j=1}^{n} Y_j/n$$

and

$$T^2 = \frac{1}{2}\left[\frac{1}{m}\sum_{i=1}^{m}(X_i - \overline{X})^2 + \frac{1}{n}\sum_{j=1}^{n}(Y_j - \overline{Y})^2\right].$$

A natural estimator for $P(X < Y)$ is then

$$\widehat{P}(X < Y) = \Phi\left(\frac{\overline{Y} - \overline{X}}{\sqrt{2}\,T}\right).$$

Show that $\widehat{P}(X < Y)$ is a consistent estimator of $P(X < Y)$ as $m \to \infty$ and $n \to \infty$ (i.e., that $\widehat{P}(X < Y) \xrightarrow{P} P(X < Y)$ as $m \to \infty$ and $n \to \infty$).

8.25. Let $X_{(1)} \leq \cdots \leq X_{(n)}$ be the order statistics for a random sample of size n from the continuous distribution with p.d.f.

$$f_X(x) = \frac{1}{2\theta}\,I_{(-\theta,\theta)}(x)I_{(0,\infty)}(\theta).$$

(a) Find the p.d.f. for $X_{(n)}$.

(b) Obtain $E[X_{(n)}]$ and $Var(X_{(n)})$.

(c) Show that $\widehat{\theta}_1 = X_{(n)} \xrightarrow{P} \theta$ as $n \to \infty$.

(d) Show that $V = -X_{(1)}$ has the same probability distribution as $X_{(n)}$.

(e) Argue that $\widehat{\theta}_2 = \frac{X_{(n)} - X_{(1)}}{2} \xrightarrow{P} \theta$ as $n \to \infty$.

(f) Set $Y_i = |X_i|$, for $i = 1, \ldots, n$, and let $Y_{(1)} \leq \cdots \leq Y_{(n)}$ be the order statistics for the Y's. Argue that $Y_{(n)} = max\,\{-X_{(1)}, X_{(n)}\}$ and then use this fact to show that $\widehat{\theta}_3 = Y_{(n)} \xrightarrow{P} \theta$ as $n \to \infty$.

Think About It We have seen that $\widehat{\theta}_1$, $\widehat{\theta}_2$, and $\widehat{\theta}_3$ are all three consistent estimators for θ. Which of these three would you prefer to use to estimate θ and how do you support your choice?

8.26. Let $X_{(1)} \leq \cdots \leq X_{(n)}$ be the order statistics for a random sample of size n from the continuous distribution with p.d.f.

$$f_X(x) = \frac{30\theta^4}{x^4} I_{(\theta,\infty)}(x)I_{(0,\infty)}(\theta).$$

(a) What is the probability distribution for $X_{(1)}$?
(b) Find $E[X_{(1)}]$ and $Var(X_{(1)})$.
(c) Show that $X_{(1)} \xrightarrow{P} \theta$ as $n \to \infty$.

8.27. Let $X_1,\ldots,X_n$ be a random sample of size n from the continuous distribution with p.d.f.

$$f_X(x) = e^{-(x-\theta)} I_{(\theta,\infty)}(x)I_{(-\infty,\infty)}(\theta)$$

and let $X_{(1)} \leq \ldots \leq X_{(n)}$ be the corresponding order statistics.

(a) Find and identify the probability distribution of $X_{(1)}$.
(b) Find $E[X_{(1)}]$ and $Var(X_{(1)})$.
(c) Show that $\widehat{\theta}_1 = X_{(1)} \xrightarrow{P} \theta$ as $n \to \infty$.
(d) What is $E[X_1]$?
(e) Argue that $\widehat{\theta}_2 = \overline{X} - 1 \xrightarrow{P} \theta$ as $n \to \infty$, where $\overline{X} = \frac{1}{n}\sum_{i=1}^{n} X_i$ is the sample mean.
(f) From parts (c) and (e), we see that both $\widehat{\theta}_1$ and $\widehat{\theta}_2$ are consistent estimators for θ. Which would you prefer to use and why?

8.28. Let $\overline{X}$ be the sample average for a random sample of size n from a probability distribution with mean μ, finite variance σ^2, and moment generating function $M_X(t)$.

(a) Find the general form of the moment generating function for $\overline{X}$, say $Q_n(t)$, as a function of $M_X(t)$.
(b) Use the result from (a) and a Taylor series expansion of $M_X(t)$ about $t = 0$ to prove the Weak Law of Large Numbers.

8.29. Let $X_1,\ldots,X_n$ be a random sample from a distribution with c.d.f. $F_X(x)$. The empirical (or sample) c.d.f. for these observations is defined by

$$F_n(x) = \frac{\#X's \leq x}{n} \quad \text{for all } x.$$

Let Y denote an additional random variable with c.d.f. $F_X(x)$ that is independent of $X_1,\ldots,X_n$. Set $W = F_X(Y)$ and $V = F_n(Y)$.

(a) Describe the components of randomness for both W and V.
(b) What is the probability distribution for W? Find $E[W]$ and $Var(W)$.
(c) Use the results from part (b) and those obtained in part (a) of Exercise 8.16 to find $E[V]$ and $Var(V)$. Compare these results with those from part (b).
(d) Obtain the probability distribution for V, and compare and contrast it with the probability distribution for W discussed in part (b).

8.30. Suppose that the random variable X has p.d.f.

$$f_X(x) = \frac{1}{2\sigma} \phi\left(\frac{x}{\sigma}\right) + \frac{1}{2\sigma} \phi\left(\frac{x-1}{2\sigma}\right),$$

where $\phi(z)$ is the p.d.f. for the standard normal distribution. Show that

$$X \xrightarrow{d} Y \sim Bernoulli\ (0.5), \quad \text{as } \sigma \to 0.$$

8.31. Consider the sequence of mutually independent random variables $X_1, \ldots, X_n$, where $X_i \sim Poisson\ \left(\frac{\mu}{i}\right)$, for $i = 1, \ldots, n$. One natural estimator (called the *maximum likelihood estimator*) for μ is

$$\widehat{\mu}_n = \frac{\sum\limits_{i=1}^{n} X_i}{\sum\limits_{i=1}^{n} \left(\frac{1}{i}\right)}.$$

(a) Show that

$$\widehat{\mu}_n \xrightarrow{P} \mu \quad \text{as } n \to \infty.$$

[You may use without proof the fact that the series $\sum\limits_{i=1}^{n} \left(\frac{1}{i}\right)$ diverges ($\to\infty$) as $n \to \infty$]

(b) Does the sequence of random variables $\widehat{\mu}_n$ also converge in distribution to μ as $n \to \infty$? Justify your answer.

8.32. Let $X_1, \ldots, X_n$ be a random sample from the *Poisson* (λ) distribution, $\lambda > 0$, and let $\overline{X}_n$ be the sample mean. Let $g(x) = x^\gamma$ for $x \geq 0$ and $0 < \gamma < \infty$.

(a) Show that

$$\sqrt{n}\ \left(g(\overline{X}_n) - g(\lambda)\right) \xrightarrow{d} V \sim n\ \left(0, v^2\right) \quad \text{as } n \to \infty.$$

Express v^2 as a function of λ and γ.

(b) For what value of γ and hence choice of $g(x) = x^\gamma$ is v^2 constant in λ (i.e., does not depend on λ)? [We call this $g(\cdot)$ the *variance stabilizing transformation*.]

8.33. Let $X_1, \ldots, X_{2m+1}$ be a random sample of size $2m + 1$ from the Bernoulli distribution with probability of success $p = \frac{1}{2}$, where m is a positive integer. Let $X_{(m+1)}$ be the sample median.

(a) Characterize the event $\{X_{(m+1)} = 0\}$ in terms of an event involving $\sum\limits_{i=1}^{2m+1} X_i$.

(b) Find a large sample ($m \to \infty$) approximation for $P(X_{(m+1)} = 0)$.

8.34. Let $X_1,\ldots,X_n$ be a random sample from the $Unif(0,\theta)$ distribution, with $\theta > 0$. Two competing estimators for θ are the *maximum likelihood estimator* $\widehat{\theta}_1 = X_{(n)}$ and the *method of moments estimator* $\widehat{\theta}_2 = \frac{2}{n}\sum_{i=1}^{n}X_i$.

(a) Show that both $\widehat{\theta}_1$ and $\widehat{\theta}_2$ are consistent estimators of θ.

(b) Show that

$$\frac{n\left(\theta - \widehat{\theta}_1\right)}{\theta} \xrightarrow{d} Exp(1) \quad \text{as } n \to \infty.$$

(c) Show that

$$\frac{\sqrt{3n}\left(\widehat{\theta}_2 - \theta\right)}{\theta} \xrightarrow{d} n(0,1) \quad \text{as } n \to \infty.$$

(d) Find the limiting ($n \to \infty$) distribution of $\sqrt{n}\left(\ln\left(\widehat{\theta}_2\right) - \ln(\theta)\right)$.

8.35. Let $X_1,\ldots,X_n$ be a random sample from the $n(\theta,1)$ distribution and set $A_n = \frac{1}{n}\sum_{i=1}^{n}|X_i|$.

(a) Show that there exists a real number μ such that $A_n \xrightarrow{P} \mu$ as $n \to \infty$.

(b) Find the limiting ($n \to \infty$) distribution of $\sqrt{n}(A_n - \mu)$.

8.36. Let $X_1,\ldots,X_n$ be a random sample from the $n(\theta,1)$ distribution, with $\theta > 0$, and set $B_n = \left|\frac{1}{n}\sum_{i=1}^{n}X_i\right|$. Show that there exist constants γ and σ such that

$$\sqrt{n}(B_n - \gamma) \xrightarrow{d} n(0,\sigma^2) \quad \text{as } n \to \infty.$$

8.37. Let $X_1,\ldots,X_n$ be independent random variables such that

$$X_i \sim f_{X_i}(x_i) = \frac{1}{\beta t_i}\, e^{\left(-\frac{x_i}{\beta t_i}\right)} I_{(0,\infty)}(x_i),$$

where each t_i is a known constant and $\beta > 0$ is an unknown parameter. The *maximum likelihood estimator* for β is $\widehat{\beta} = \frac{1}{n}\sum_{i=1}^{n}\left(\frac{x_i}{t_i}\right)$.

(a) Show that $\widehat{\beta}$ is an unbiased estimator for β.

(b) Find the limiting ($n \to \infty$) distribution of $\sqrt{n}\left(\widehat{\beta} - \beta\right)$.

8.38. Let $X_1,\ldots,X_n$ be a random sample from the *Geom* (p) distribution, with $0 < p < 1$. One common statistic (the *maximum likelihood estimator*) used to estimate the parameter p is $\widehat{p} = \frac{1}{\overline{X}}$, where $\overline{X}$ is the sample mean.

(a) Find the limiting ($n \to \infty$) distribution of $\sqrt{n}(\hat{p} - p)$.
(b) What are the limits of the asymptotic variance of $\hat{p}$ as $p \to 0$ and as $p \to 1$?
    Find the value of p that maximizes the asymptotic variance of $\hat{p}$.

8.39. Let $Y_n \sim NegBin\left(r, \frac{\lambda}{n}\right)$, where r is a positive integer, $\lambda > 0$, and n is any integer such that $n > \lambda$. Use the moment generating function approach to show that

$$\frac{Y_n}{n} \overset{d}{\to} Y \sim Gamma\left(r, \frac{1}{\lambda}\right) \quad \text{as } n \to \infty.$$

8.40. Let $Z_n \sim Binom\left(n, \frac{\lambda}{n}\right)$, where $\lambda > 0$ and n is any integer such that $n > \lambda$. Use the moment generating function approach to show that

$$Z_n \overset{d}{\to} Poisson(\lambda) \quad \text{as } n \to \infty.$$

8.41. Let $X_1, \ldots, X_n$ be a random sample from the continuous distribution with p.d.f.

$$f_X(x) = \frac{1}{2}\,(\pi x)^{-\frac{1}{2}}\,\lambda\,e^{-\lambda\,x^{\frac{1}{2}}\pi^{-\frac{1}{2}}}\,I_{(0,\infty)}(x)\,I_{(0,\infty)}(\lambda).$$

One statistic used to estimate λ (the *maximum likelihood estimator*) is

$$\hat{\lambda} = n\,\pi^{\frac{1}{2}}\left(\sum_{i=1}^{n} X_i^{\frac{1}{2}}\right)^{-1}.$$

Find the limiting ($n \to \infty$) distribution of $\sqrt{n}\left(\hat{\lambda} - \lambda\right)$.

8.42. Let $X_1, \ldots, X_n$ be a random sample from the *Poisson* (λ) distribution with $\lambda > 0$. Suppose we are interested in estimating $p = P(X_1 = 0) = e^{-\lambda}$. One estimator for p (the *maximum likelihood estimator*) is $\hat{p} = e^{-\overline{X}}$, where $\overline{X}$ is the sample mean.

(a) Show that $\hat{p} \overset{p}{\to} p$ as $n \to \infty$.
(b) Find the limiting ($n \to \infty$) distribution of $\sqrt{n}(\hat{p} - p)$.
(c) Is $\hat{p}$ an unbiased estimator for p? Justify your answer.

8.43. Let $X_{(1)} \leq \ldots \leq X_{(n)}$ be the order statistics for a random sample of size n from the *Exp* (1) distribution. Show that

$$Z_n = X_{(n)} - \ln(n) \overset{d}{\to} Z \quad \text{as } n \to \infty,$$

where Z follows the (Type I) Extreme Value Distribution with c.d.f.

$$F_Z(z) = e^{-e^{-z}}\,I_{(-\infty,\infty)}(z).$$

8.44. Let $X_1, \ldots, X_n$ be a random sample from the continuous distribution with p.d.f.

$$f_X(x) = \theta \, (1+x)^{-(1+\theta)} \, I_{(0,\infty)}(x) I_{(2,\infty)}(\theta).$$

One statistic (*method of moments estimator*) that can be used to estimate θ is $\widehat{\theta} = \frac{\overline{X}+1}{\overline{X}}$.

(a) Find $E[X_1]$ and $Var(X_1)$.

(b) Show that $\widehat{\theta} \overset{P}{\rightarrow} \theta$ as $n \to \infty$.

(c) Find the limiting ($n \to \infty$) distribution of $\sqrt{n}\left(\widehat{\theta} - \theta\right)$.

(d) Find the limiting ($n \to \infty$) distribution of $W_n = \dfrac{\sqrt{n}\left(\widehat{\theta}-\theta\right)}{\left(\widehat{\theta}-1\right)\sqrt{\dfrac{\widehat{\theta}}{\widehat{\theta}-2}}}$.

8.45. Let $X_1, \ldots, X_n$ be a random sample from a probability distribution with mean $\mu = \lambda$ and variance $\sigma^2 = \lambda$, and let $\overline{X} = \frac{1}{n} \sum_{i=1}^{n} X_i$ denote the sample mean.

(a) Find the limiting ($n \to \infty$) distribution of $V_n = \dfrac{\sqrt{n}(\overline{X}-\lambda)}{\sqrt{\overline{X}}}$.

(b) Describe a discrete probability distribution that satisfies the conditions of this exercise.

(c) Describe a continuous probability distribution that satisfies the conditions of this exercise.

8.46. Let $X_1, \ldots, X_n$ be a random sample from the Gamma (α, β) distribution with $\alpha > 0$ and $\beta > 0$, and let $\overline{X} = \frac{1}{n} \sum_{i=1}^{n} X_i$ and $S^2 = \frac{1}{n} \sum_{i=1}^{n} (X_i - \overline{X})^2$ be the sample mean and sample variance, respectively.

(a) Find the limiting ($n \to \infty$) distribution of

$$Q_n = \frac{\sqrt{n}\,(\overline{X} - \alpha\beta)}{\sqrt{\alpha}\,\beta}.$$

(b) Find the limiting ($n \to \infty$) distribution of $T_n = \frac{\sqrt{n}\,(\overline{X}-\alpha\beta)}{S}$.

8.47. Let $X_1, \ldots, X_n$ be a random sample from the *Unif* $(0, \theta)$ distribution with $\theta > 0$, and let $\overline{X} = \frac{1}{n} \sum_{i=1}^{n} X_i$ be the sample mean. Find the limiting ($n \to \infty$) distribution of $Q_n = \frac{\sqrt{3n}(2\overline{X}-\theta)}{2\overline{X}}$.

8.48. Let $X_1, \ldots, X_n$ be a random sample from the *Binom* (m, p) distribution, where m is a positive integer and $0 < p < 1$.

(a) Find an expression for $\gamma = P(X_1 = 0)$ as a function of m and p.

(b) Let $T_n = [\#X_i's = 0]$. What is the probability distribution for T_n?

(c) Let $\widehat{\gamma} = \frac{T_n}{n}$ be the sample percentage of X's that are zero. What is the limiting $(n \to \infty)$ distribution of $W_n = \frac{\sqrt{n}\left(\widehat{\gamma}-\gamma\right)}{\sqrt{\widehat{\gamma}\left(1-\widehat{\gamma}\right)}}$? Justify your answer.

(d) Let $\overline{X} = \frac{1}{n} \sum_{i=1}^{n} X_i$ be the sample mean. Find the limiting $(n \to \infty)$ distribution of

$$V_n = \frac{\sqrt{n}\left(\widehat{\gamma}-\gamma\right)}{\sqrt{\left(1-\overline{X}\right)^m \left[1-\left(1-\overline{X}\right)^m\right]}}.$$

(e) Which of the two asymptotic results in parts (c) and (d) do you think would be better in terms of approximating the properties of $\widehat{\gamma}$ for large n? Why?

8.49. Let $X_1, \ldots, X_n$ be a random sample from the *Poisson* (λ) distribution with $\lambda > 0$.

(a) Find an expression for $\delta = P(X_1 \leq 0)$ as a function of λ.

(b) Let $Q_n = \left[\#X_i's \leq 0\right]$. What is the probability distribution for Q_n?

(c) Let $\widehat{\gamma} = \frac{Q_n}{n}$ be the sample percentage of X's that are either 0 or 1. What is the limiting $(n \to \infty)$ distribution of $W_n = \frac{\sqrt{n}\left(\widehat{\gamma}-\gamma\right)}{\sqrt{\widehat{\gamma}\left(1-\widehat{\gamma}\right)}}$? Justify your answer.

(d) Let $\overline{X} = \frac{1}{n} \sum_{i=1}^{n} X_i$ be the sample mean. Find the limiting $(n \to \infty)$ distribution of

$$V_n = \frac{\sqrt{n}\left(\widehat{\gamma}-\gamma\right)}{\sqrt{\left(1+\overline{X}\right)e^{-\overline{X}}}}.$$

(e) Which of the two asymptotic results in parts (c) and (d) do you think would be better in terms of approximating the properties of $\widehat{\gamma}$ for large n? Why?

8.50. Let $X_1, \ldots, X_n$ be a random sample from the *Beta* (α, α) distribution with $\alpha > 0$. Let $\overline{X} = \frac{1}{n} \sum_{i=1}^{n} X_i$ and $S^2 = \frac{1}{n} \sum_{i=1}^{n} \left(X_i - \overline{X}\right)^2$ be the sample mean and sample variance, respectively.

(a) Find the limiting $(n \to \infty)$ distribution of $W_n = 2\sqrt{2\alpha+1}\ \sqrt{n}\left(\overline{X} - \frac{1}{2}\right)$.

(b) Find the limiting $(n \to \infty)$ distribution of $V_n = \sqrt{n}\frac{\left(\overline{X}-\frac{1}{2}\right)}{S}$.

Bibliography

AAA (2016) Foundation for traffic safety: prevalence of self-reported aggressive driving behavior: United States, 2014. Report Issued July 2016. www.aaafoundation.org

Business Insider (2012) Jay Yarrow, BUSINESS INSIDER, 30 Aug 2012. 51% of people think stormy weather affects 'cloud computing'. http://www.businessinsider.com/. Accessed 21 June 2016

Cable News Network (CNN) (2009) Survey: support for terror suspect torture differs among the faithful. 30 Apr 2009. http://edition.cnn.com/2009/US/04/30/religion.torture

Cramér H (1946) Mathematical Methods of Statistics Princeton University Press, Princeton, New Jersey

GfK Mediamark Research & Intelligence, LLC (2015) 44% of US adults live in households with cell phones but no landlines. Press release, 2 Apr 2015. New York. www.gfk.com/us

Meilman PW, Leichliter JS, Presley CA (1998) Analysis of weapon carrying among college students, by region and institution type. J Am Coll Health 46(6):291–299

National Public Radio (2014) Scott Neuman, 14 Feb 2014. http://www.npr.org/sections/the two-way/2014/02/14/277058739/. Accessed 21 June 2016

Pew Research Center (2010) Social and demographic trends: women, men and the new economics of marriage, by Richard Fry. Report issued 19 Jan 2010. www.pewresearch.org

Udias A, Rice J (1975) Statistical analysis of microearthquake activity near San Andreas Geophysical Observatory, Hollister, California. Bull Seismol Soc Am 65:809–828

USA Today (2013) Chris Chase, USA TODAY Sports. 30 Jan 2013. http://www/usatoday.com/story/gameon/2013/01/30. Accessed 21 June 2016

© Springer Nature Switzerland AG 2020
D. Wolfe, G. Schneider, *Primer for Data Analytics and Graduate Study in Statistics*, https://doi.org/10.1007/978-3-030-47479-9

Printed in the United States
by Baker & Taylor Publisher Services